[illegible] the Grain

The Genetic Transformation of Global Agriculture

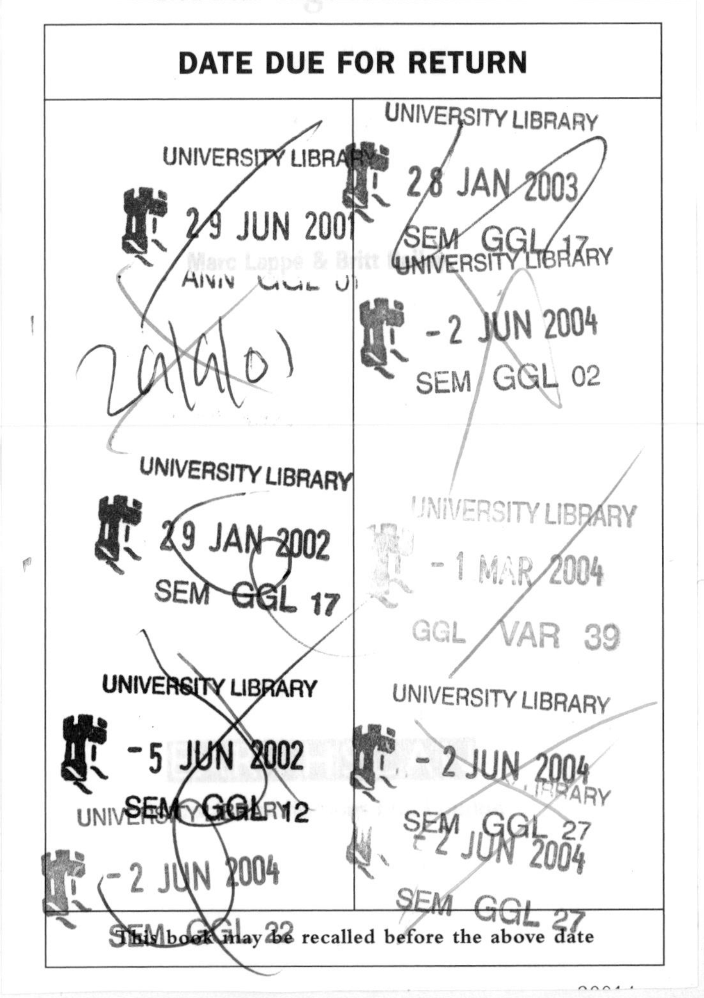

Dedication

Dedicated to Frances Moore Lappé and the memory of Judi Bari and Rachel Carson, three champions of right stewardship

First published in the UK in 1999 by
Earthscan Publications Limited

Reprinted 1999 (hardback)

A catalogue record for this book is available from the British Library

ISBN: 1 85383 576 5 (hardback)
1 85383 657 5 (paperback)

Design by Studio 2D, Champaign, Il
Printed and bound by Biddles Ltd, Guildford and King's Lynn
Cover design by John Burke Design (hardback)
Cover photo © Graham Burns/Environmental Images (hardback)

For a full list of publications, please contact:

Earthscan Publications Ltd
120 Pentonville Road
London N1 9JN
Tel: (0171) 278 0433
Fax: (0171) 278 1142
email: earthinfo@earthscan.co.uk
http://www.earthscan.co.uk

Earthscan is an editorially independent subsidiary of Kogan Page Limited and publishes in association with WWF-UK and the International Institute for Environment and Development.

This book is printed on elemental chlorine free paper.

Contents

Foreword

Tim Lang

There can be few people in the world who are not, if only dimly, aware that something significant is going on down on the farm. The words "biotechnology" and "genetic engineering" are now part of the lexicon. These words are not the usual froth of languages in transition, words which are one moment fashionable and imbued with meaning, the next moment archaic and no longer "cool". Genetic engineering is one of those seismic shifts in human life that we usually read about in history books as having happened, yet this is one we are now part of. The invention of the wheel, the internal combustion engine, the computer, for example, have all shaped our lives to an incalculable extent. The difference with this one is that we are not just observing or relating with this technological revolution, but quite literally eating it. The biotechnology revolution is shattering our conceptions of life and nature in a very personal, immediate way. No wonder so many people find this so troubling. No wonder everywhere in the world, debate is so vibrant. The issue, as this book reminds us, is immense.

Whatever our position on genetic engineering – supporter, critic or sceptic – we have to recognize the extraordinary feat that it represents. Life is being refashioned. It is humbling to see how fast this is being done and to note the range of foods to which the technology is being applied. Like them or not, brilliant advances in the science of life are now being applied to our food. This book is well worth reading just to appreciate that process, but just because we can acknowledge the creativity and wizardry of the scientists does not mean that we should be seduced by them. Rightly, the huge moral questions raised are now being debated but, in my view, not enough attention is being given to a hard business element behind the rapid arrival of biotech products on supermarket shelves. This is where this book is particularly welcome.

Most of the huge investments in biotechnology and genetic manipulation have been made over decades by governments and private companies, not by individual scientists working alone in dusty garret laboratories. This is big commerce science. One might argue that there is nothing unusual about this in the modern world where giant corporations dominate our lives as consumers, workers, citizens, domestic beings. If a large corporation is prepared to stake its reputation on the frontiers of science, surely is that not to our general benefit? It is at this point that my own doubts become serious. The great advances that technologies have been able to make to our quality of life, from the humble domestic vacuum cleaner to hi-tech commodities such as the television, have all come to market through decades of experimentation. When they penetrate our lives deeply, it is an herculean task to try to control them. Look at the car. Seeing the way this machine has come to dominate our lives – we go to work to earn money to be able to afford a car, to use it to escape the lives we have created, and all the while say "isn't it convenient?"! – it is entirely legitimate to question genetic engineering now before it really is too late. Some argue that the point of no return is already passed, but I do not. Either consumers have power and the right to have choice or they don't; and if they don't, one of the West's great ideological building blocks would be exposed as a sham with terrible consequences.

By contrast with the arrival of the car or the wheel, the pace and scale of biotechnology's arrival is remarkable. Whom do we have to justify this? Why, the scientists and companies themselves. Hurry, hurry, hurry, they have been urging European legislators. If you do not allow us to pursue the biotech revolution, Europe will be left standing while the US takes all. Invest, invest, invest, they tell the stockmarket. This is the new oil rush. Bonanza time. Invest heavily now or you will not be able to keep up with the rapid developments in the science. These are seductive arguments for all who are not already clear where they stand. It is on this point that Marc Lappé and Britt Bailey are particularly useful. As US researchers, they are well placed to explore the US corporations who are driving genetic engineering in food. While other companies in other countries are also big players, there is little doubt that the US is the epicentre of the new movement. Only it has had the right combination of corporate coffers deep enough to make the necessary investments, ample stockmarkets, entrepreneurialism, armies of suitably qualified scientists and farmers locked onto an intensification treadmill, to be prepared to risk planting the products. Lappé and Bailey's narrative is a wonderful mix of interviews, reportage and factual analysis that allows the reader to appreciate this blend.

Although many commentators argue that the future is clear and that biotechnology is here to stay, I think it is not clear what will happen in this latest food revolution. By their nature, revolutions are almost unstable, unpredictable. The big players in this field, the companies, are indeed desperately trying to exert control, to make events predictable, to set liquid into social concrete, to get their way. But there is a little structural difficulty en route. People have got to eat these products. This is why there has been such a lengthy and fraught battle in the European Union about regulations and laws. This is set to continue in the near future, but what about the longer term?

If we characterize opinion on biotechnology in food as a spectrum, at each end there are predictions of total change. The critics prophesy environmental hazards and possible problems for human health. The proponents announce a cornucopia, the end of starvation and the dream of farming being able to control its products at last. Who knows what will happen? Pending the outcome, I detect at least three key tensions on which the actions of the players – the proponents, opponents, undecided – will make a significant difference. These are the fault-lines which will characterize the territory for the future.

The first is the science itself. Thus far in the debate, most pressure has come from outside the scientific world. Rapidly, people with strong intellectual stakes within science are themselves beginning to express doubts. I think, but I may be wrong, I detect the beginning of the end of the research bonanza in its current twenty year old form. Until now few scientists in the biological or life sciences area (as they now like to call it) could resist the lure of the big grants and commercial sponsorship. Empires and careers have been made on it, heralding the 21st century as the biological century. But after the BSE (mad cow) crisis, the scientific community, in Europe particularly, has been more open in thinking the unthinkable, asking moral questions *before* banking the grant. They are right to see the tension over biotechnology in food as not just a battle over science but a battle within science. They are right to hesitate before prophesying too much. An era of hard slog rather than hype within and about science is upon us.

The second tension is within food culture. If the 20th century can be characterized as the era of the industrialization of food, the 21st century may take on a different orientation. I see very contrary trends already. On the one hand, there are immense pressures to continue intensification and industrialization of food systems from farm to plate. Clearly, biotechnology fits this trend. On the other hand, I also detect strong moves in another direction; not towards some fictional "natural" food (that stopped when settled agriculture began 8–10,000 years ago) but towards more local, diverse and less processed foods. In this respect, the question to ask is not whether genetically engineered foods may exist but whether people will eat them? This is why the big corporations are now pouring huge sums into public relations budgets. They fear losing sales after their billions of dollars of investment. They are desperate to win the battle of hearts and minds because they now realize that mass psychology is as important as biology to the investment plans.

The final tension is commercial. Is genetic engineering a good investment? I happen to agree with Lappé and Bailey that genetic engineering is "an extension of corporate dominance" (p16), but you the reader or your mother, may not. She may be happy to take the dividend from her pension investment in the company (while perhaps not eating their product). It is here that a real minefield for the companies opens up. Essentially, many have moved into the area as an extension of agrichemicals or as an escape route from them. But what if biotechnology does not work out quite as its strongest proponents argue? A sensible investment strategy would be to hedge one's bets. *Not* to put all your money within a corporation on the biotech horse. Over the years, agrichemical companies have vertically integrated with seeds and sales. Monsanto, as I write, is on another spending spree buying up seed companies. One way for the biotech companies to hedge their bets is the classic route of buying the competition and the outlets. But is this a good investment for the pension funds to be in? I suspect the hard finance questions will become more salient in coming years. Shares can go up, but also down.

This book is most welcome. It takes us through the technology, introduces us to key players, humanizes the processes and does it all in a most accessible and readable manner. We need more such books to increase our cultural literacy. The issues are with us and will surely not go away.

Tim Lang PhD
Professor of Food Policy
Centre for Food Policy
Thames Valley University

Foreword to the American Edition

J. B. Neilands

There is a maxim that proclaims: "If you eat, then you are involved with agriculture." Unfortunately, the vast majority of the people on this planet appear to be oblivious to the quiet revolution now well underway in the methodology of food production. While most of us have been made aware of the widespread application of recombinant DNA technology in the biomedical field, few seem to appreciate that this powerful technology is also being applied in agriculture. The authors mention that little on the subject has appeared in *Time, Newsweek,* and *USA Today.* It has remained for publications like the *Anderson Valley Advertiser, Slingshot, Food & Water* and similar media with very modest circulation to alert people to what is actually going on. *Against the Grain* will help to correct this imbalance.

This book is primarily an exercise in investigative journalism. Not content to draw their conclusion from the glossy brochures of agribusiness, Dr. Lappé and Ms. Bailey have gone out in the field and made firsthand inspections of the supposed benefits of transgenic plants. What you hold in your hand is the impassioned report of two individuals well-versed in plant science, genetics, and ethics. Their contention, thoroughly documented with facts and citations, is that the quest for corporate profits has ridden roughshod over questions of public health, freedom of choice and ecological stability.

Against the Grain outlines the basic mechanics of manipulating DNA, starting with the classical work of Griffith and Avery, the first and largely unrecognized scientists to elucidate the genetic blueprint of life. DNA, individual segments of which comprise a gene, code for synthesis of proteins, such as enzymes, which do the actual work of the cell. How technically "sweet" it would be if foreign genes could be inserted into plants to give the recipient characters which we humans deem to be desirable. In the beginning, as explained in the book, only the large tumor-inducing (Ti) plasmid of *Agrobacterium tumefaciens* was available as a vector. This had limitations as a general carrier and has since been superseded by agents which can transform monocots, the source of grains and much of the world food supply for humans and livestock.

Dr. Lappé and Ms. Bailey focus on three systems—two herbicides and one toxin—to which transgenic technology has been applied. These are the herbicides bromoxynil and Roundup® (glyphosate), and the proteinaceous toxin of *Bacillus thuringiensis (Bt)*. Bromoxynil, a product of Rhöne-Poulenc, is a simple organic molecule with a toxic nitrile function. An enzyme known as nitrilase destroys the toxicity (for plants) by the successive addition of two moles of water, and together with other transformations leads to 3,5-dibromo-4-hydroxybenzoic acid (DBHA) as the end product. The gene for nitrilase has been inserted in various crops but, according to the authors, the animal toxicity of DBHA has never been ascertained. The toxicity of Roundup®, which inhibits an enzyme making aromatic amino acid in plants, appears to be low. However, the quantities Monsanto is planning to apply to their proprietary Roundup Ready™ cultivars, modified to carry the gene for the target enzyme, humbles the imagination. Finally, *Bt* toxin, also a Monsanto product, is designed to kill insects feeding on leaves programmed with the gene for the foreign protein. This is almost certain to generate resistant strains and to render *Bt* toxin useless to organic farmers. These are only some of the problems cited. Similar critiques could be leveled against other strategies, such as the use of anti-sense technology to delay ripening.

The giant chemical firms—Hoechst, Rhône-Poulenc, Ciba, Sandoz (Novartis), and Monsanto—have all jumped on the bioscience bandwagon, causing the *Chemical and Engineering Newsletter* of the American Chemical Society, to ask in its October 6, 1997, issue: "Who's going to make chemicals?" A three page advertisement in *The New York Times* for August 19, 1997, lays out the strategy adopted by Monsanto. The first page carries two sentences in the middle of a blank sheet: "Let's see now. What's a simple way for business people to understand the difference between the two companies Monsanto is becoming?" The second page is boldly titled "Stalks" and explains that the company is committed "to provide better food, better nutrition, and better health." This page is linked to the third by a large ampersand positioned directly over the centerfold. That third page is titled "Bonds." The text announces the new name, Solutia, and gives the address on the New York Stock Exchange. Clearly, the motivation is profit rather than nutritional quality or yield.

What is to be done? I believe that labeling would be in order for all transgenically modified foods. But even this "liberal" gesture is denounced by agribusiness which, in fact, is promoting what they call "food disparagement" laws. Reading *Against the Grain* will afford some immunity against the flummery of company detail men. It will also assure some independence from the academic high priests and whiz-kids of molecular biology who may very well be trained with conflicts of interest. If feasible, grow your own. Farmer's markets should be patronized, bearing in mind that transgenic seeds may soon be the sole type available. Ultimately, it may be necessary to resort to an independent laboratory analysis of the genome of staple foods. Data from agribusiness and government labs should be viewed with caution. Private labs may be considered credible only if they maintain economic independence from the agribusiness complex.

A hasty genetic transformation of world food resources ignores the wisdom enshrined in eons of original evolution. *Against the Grain* gives abundant reasons for health, environmental and ethical concerns, and predicts that the world may be facing a disaster of epic dimension.

J. B. Neilands
Professor Emeritus
Department of Biochemistry and Molecular Biology
University of California, Berkeley

Acknowledgments

We are extremely grateful to have received the cooperation and support of a large number of people in the farming, manufacturing and technology sector. We especially wish to thank Mr. George Zanone, of Horseshoe Lake, Arkansas, and Mr. Lowell Taylor of Hughes, Arkansas, who gave generously of their time to explain their respective transgenic and conventional farming methods. We are grateful for the technical assistance of Anthony Lappé who videotaped our research trip to Arkansas and Missouri. We also benefitted immensely from the responsiveness of many of the major corporations whose representatives spoke with us about their use policies for genetically engineered food crops.

We are extremely grateful for the help and assistance of Hartz Seed Company for permitting us to videotape and interview their staff in Stuttgart, Arkansas, and for providing us with samples of genetically engineered seeds. We thank the Arkansas Cooperative Extension Service of the University of Arkansas for invaluable aid and assistance, especially Dr. Ford Baldwin, Lanny Ashlock and Bill Robertson, who provided invaluable information about the crop yields and performance of genetically engineered seed.

We are also grateful to our Swedish colleague, Martin Frid, who provided us with much information and data about the European experience, Jane Rissler at the Union of Concerned Scientists, and Michael Gregory, Director of Arizona Toxics Information. We also thank Dr. Norman Ellstrand from University of California at Riverside for agreeing to an interview which provided key data about pollen distribution and genetic diversity.

We extend our gratitude to our book designer, Gretchen Wieshuber, for her creativity and diligence in bringing this project to fruition. We are also grateful to Ann Mitchell Kelly who contributed the graphs and charts for the text.

We thank Professor Joe Neilands of the University of California, Berkeley, who put aside much of his own work to dive into a foreword for the book. We also appreciate the advice and support of colleagues, particularly Sheldon Krimsky of Tufts University who sent us key articles and reprints.

We also want to warmly thank our families, especially Carolyn and Doug Bailey who graciously put up our research team in Memphis, Tennessee. We are both warmly supported by our loved ones, C. L. and J. L. Finally, we want to warmly thank Mr. Greg Bates of Common Courage Press for *his* courage and integrity in publishing this book.

Introduction

We are on the cusp of a major revolution in the way we grow our crops, a revolution fueled by biotechnology and driven by multinational corporations. This revolution is unique because it entails the first major agricultural transformation of food crops based entirely on genetic engineering. It is also remarkable from a sociological perspective. Many of the key innovations have occurred behind academic and corporate doors with little public input. As a result, the public response in the United States has been strangely muted.

In marked contrast to the American reaction to agricultural biotechnology, the whole concept of genetically engineering food has alarmed many people in Europe, Great Britain, and Australia. A small but vocal group of Americans committed to organic, pesticide-free and genetically untampered foodstuffs has also become increasingly disturbed. Among other issues, many of these proponents of natural foods are upset by the silent intrusion of artificially created commodities into our food supply. Are such concerns simply emotionally driven, or are they undergirded by bona fide issues of scientific uncertainty, health risks, and ecological dangers?

In this book, we intend to describe and then get beyond the emotional and political concerns. We are fundamentally interested in the tension between the promise and peril of genetically engineered plants. For many crops, like soybeans and corn, genetic engineering is a *fait accompli,* and issues of health, safety, and nutritional quality are seemingly resolved. From our conversations with corporate representatives, any lingering concerns about safety or health are deemed irrational, overblown, or exaggerated. Corporate scientists point out—correctly—that most transgenic crops are genetically identical to their original progenitors in all but a handful of their tens of thousands of genes. What, then, is there in the construction of new organisms with genes imported from disparate species that warrants concern, you may ask?

Like many others, we are especially concerned about possible health hazards stemming from ingestion of genetically engineered food crops. In researching this issue, we have been frustrated by the lack of good science on which to base a final opinion. So much of the genetically engineered produce has been rushed to market that health effect assessments have been few and far between. What is available deals primarily with the toxicity profile of the herbicides which will come into greater play as genetically engineered crops are commercialized. As we will show, a full spectrum of toxicity can be found. Some of these herbicides, like glyphosate, appear to be relatively benign to their immediate users, while others, like bromoxynil, are quite toxic. Beyond these relatively simplistic assessments lies a disturbingly large universe of uncertainty. Few studies have been performed on ecological impacts of long-term reliance on transgenic cropping methods. And no studies exist that measure the impact of chronic ingestion of finished products containing increased amounts of transgenic crops. Most disturbingly, we may never know these consequences because transgenic food sources are presently marketed cheek-to-jowl with conventional ones. Without any labels to distinguish one from the other, epidemiologists are hamstrung. Should there be a biological effect from ingesting altered soybean by-products (as may result from changes in their estrogenic components), we presently lack any way to identify or track affected populations. As we will show, such an issue is particularly compelling for babies who may drink soy formula or children who increase their ingestion of soy products in candy or diet generally.

We can identify a group of unresolved questions:

- Are there hidden risks in moving genes across species lines in plants?
- Have we missed possible secondary effects of gene transfer across whole crop lineages in assessing the safety of genetically engineered crops?
- Is there a loss of opportunity for genuine progress in choosing certain genes for crop development and not others?

- Are novel ecological risks created from developing monocultures of genetically engineered crops?
- Are special problems of disease resistance and vulnerability posed by transgenic crop types?
- Will new patterns of pesticide use create health risks to farm workers or consumers?

While many of these questions will be answered through review of scientific studies and projections, many other issues raised by genetic engineering of food crops require political and cultural analysis.

The Scope of the Problem

If biotechnology were simply a novel scientific enterprise, it would be fitting to focus our concerns solely on its safety or environmental impacts. Such objections to genetic manipulation preceded the planting of the first genetically engineered crop by at least a decade. In the early 1970s, the first genetically engineered crop was created. The first genetically altered bacteria designed to make strawberries frost resistant were planned for release into a strawberry field in California. Environmental groups concerned about the introduction of new diseases and the possibly catastrophic spread of genetically engineered bacteria to other crops protested loudly—and effectively. The planting was postponed for two seasons. When it was finally done the results were disappointing. Modest frost protection was achieved, but no great ecological damage ensued. The first wave of opposition to genetically engineered crops was a tempest in a teapot. Neither the bacteria nor their altered genes proved to be a health or environmental hazard—both were short lived and nonpathogenic.

But more robust genes and organisms were planned for testing. The methionine-rich gene from a Brazil nut was slated to be introduced into soybeans to increase their protein level. Fortunately, before the soybeans were commercially released, a group of scientists discovered that the "new" soybeans contained the allergenic properties of the Brazil nut.[1] Some 13 different field trials of transgenic plants have been sabotaged in Germany in the belief that they posed imminent hazards to the ecosystem. Most recently, a transgenic beet crop planted in Ireland was uprooted, allegedly by activists concerned about the spread of herbicide resistant genes.

Our view, and the focus of this book, is that genetically engineered crops pose a much more insidious threat, one that promises to shift our cultural mores and agricultural practices as a whole. It is in their "success" that genetically engineered plants pose the greatest risk. By sidestepping regulatory review and public oversight, arbitrarily chosen genetically engineered plants promise to transform American and—ultimately—world agriculture.

Issues of Scale

The concerns detailed in this book are triggered as much by the scale of present use of transgenic crops as by its scientific uniqueness. In the last year alone, literally hundreds of new crops have been proposed for field testing. We document the varieties in Chapter 2. As shown in Figure 1, in the 1996-97 planting season many commodities derived from biotechnology research have been marketed and planted on over 30 million acres in the United States.[2] These include herbicide resistant field crops such as corn, soybeans, cotton and canola; insecticide resistant crops such as cotton, corn and potatoes; delayed ripening tomatoes; genetically altered soybeans with high-oleic acid oil; alkaline-tolerant corn; and virus resistant squash.

Corporate Dominance

One company in particular has dominated this emerging agriculture/biotechnology (agbiotech) market (see Figures 1, 2 and 3). Monsanto Company, based in St. Louis, Missouri, has capitalized on its glyphosate herbicide known as Roundup® by selectively engineering crops to be resistant to high levels of this herbicide. Among successfully transformed crops are sugar beets, corn, cotton, rice and soybeans. The particular genetic technique used to convert native strains or cultivars of these crops—or more commonly pre-made hybrids—to an herbicide tolerant state is known as Roundup Ready™ technology. This radical new technology, to be described below, allows plants to withstand high doses of Roundup® that would otherwise destroy them along with their weedy neighbors.

In the four years since its inception in 1994, Roundup Ready™ technology has been applied full force to American seed crops. Barely

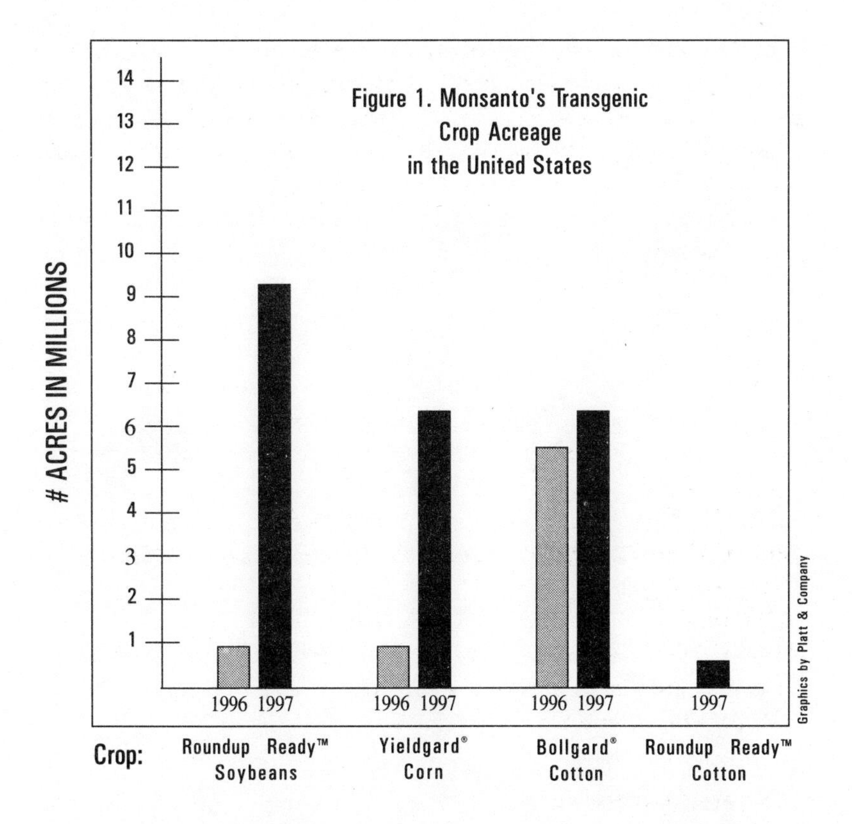

off the drafting boards in 1997, 15% of the soybean and 14% of the cotton crop was engineered with Roundup Ready™ genes. In 1998, the first corn crops are planned to be engineered with this herbicide resistant gene. According to Hartz Seed Company, a Monsanto subsidiary, Monsanto plans to have half of the United States soybean crop, or some 30 million acres, genetically engineered with Roundup Ready™ genes by 1998. According to the same source, in the year 2000, Monsanto plans to have 100% of soybeans in the United States converted with the Roundup Ready™ gene technology.[3]

An American Perspective

In the United States few people know much about the dramatic events that undergird this transformation of American agriculture. Descriptions of this revolutionary technology in popular publica-

Figure 2. Monsanto's Projected *Roundup Ready™* Crop Program

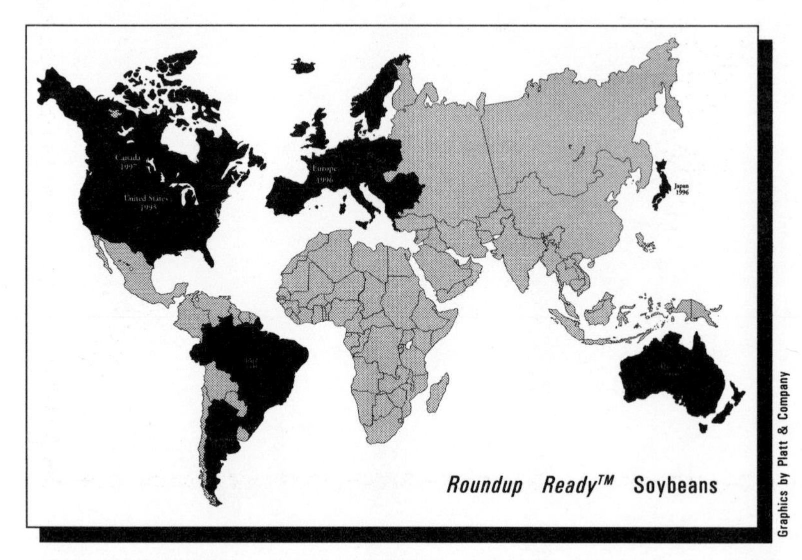

Canada 1998
United States 1996
Europe 1996
Japan 1997
Brazil 2000
Roundup Ready™ Corn

tions like *Time, Newsweek* or *USA Today* are a rarity. Such media avoidance of crop genetic engineering would be understandable if we were discussing corn or soybean futures on the commodity market or the price of a bushel of wheat. But what we are discussing

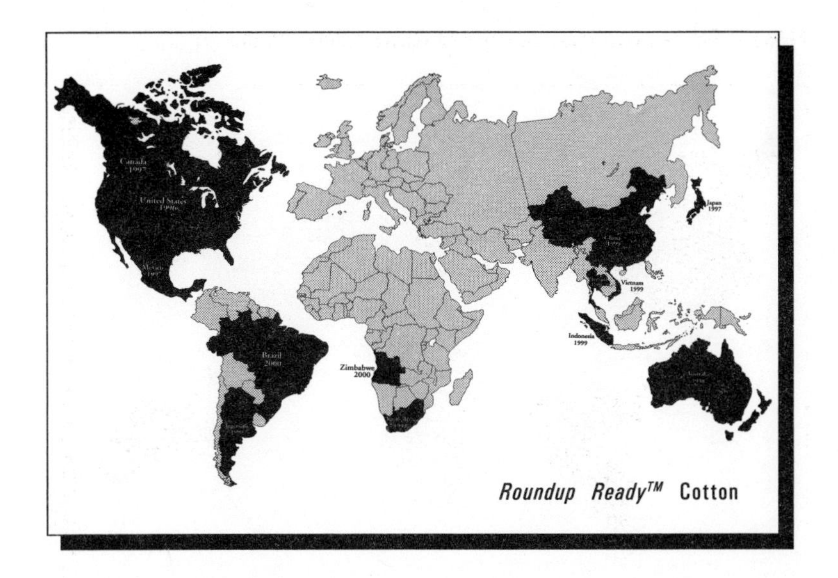

Roundup Ready™ Cotton

here promises to affect all our lives. If Monsanto and other primary actors like Dow Chemical, Novartis, and Dupont have their way, within the next four to five years, essentially all of the conventionally grown soy products we consume will be derived from genetically engineered crops. Lest you think you can avoid having to face this growing reality, consider that virtually all candy, chocolate bars, ice cream, cookies and salad dressings contain products derived from genetically engineered plants. So too will most of our meat, at least indirectly, as genetically engineered crops like soybeans enter livestock diets in increasing proportions.

Exploring Differences

Given the pending ubiquity of gene-engineered crops in our present and future diet, why has there been so little concern expressed about this transformation in the United States? In marked distinction to our passivity, many European governments are militantly opposed to the production, importation, or sale of transgenic crops. Luxembourg, Switzerland, and Denmark head up this revolution in consciousness. Each has taken special steps to ban or limit importation or use of genetically engineered crops.

Figure 3. Countries Projected to Use Monsanto's Insect-Resistant Crops

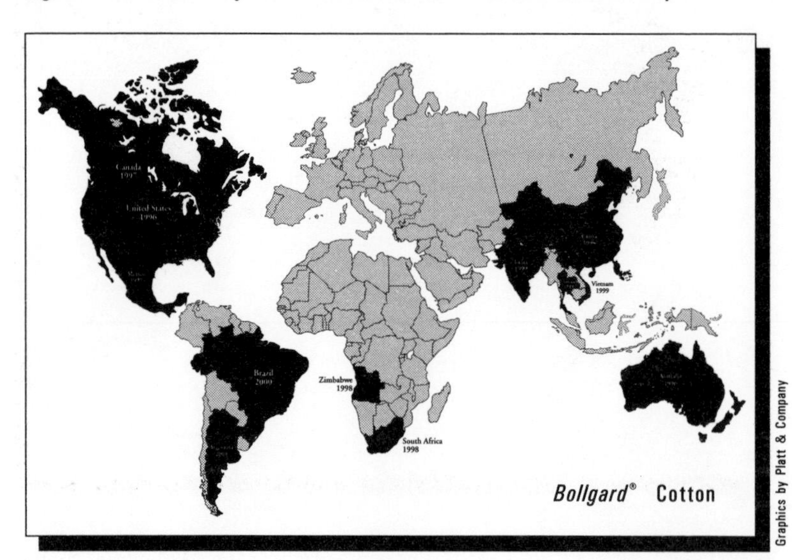

Japan 1997

Yieldgard® Corn

Graphics by Platt & Company

The Secretary of the United States Department of Agriculture (USDA), Dan Glickman, has dismissed such actions as misguided and uninformed. In a famous speech, he urged the European Union to recognize the legitimacy and safety of what he called the "Second

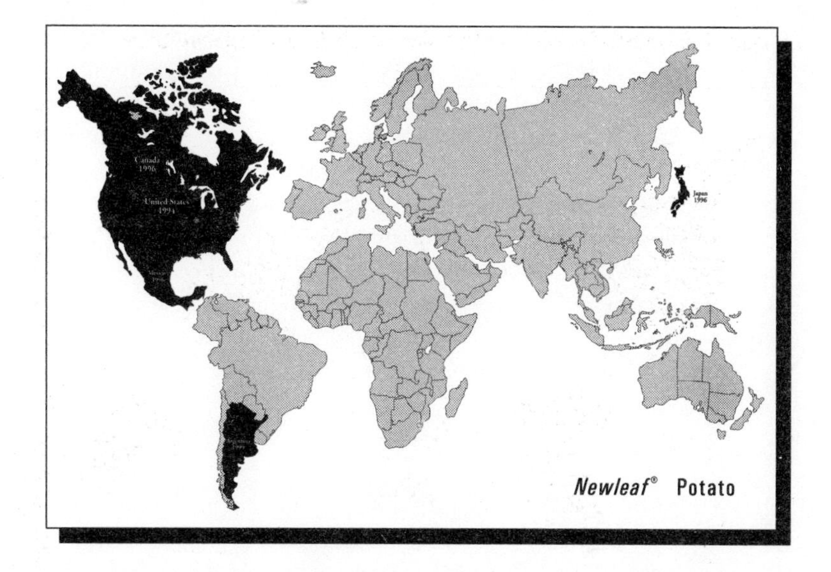

Green Revolution."[4] Glickman characterized the European reaction to genetically engineered foods as overblown, culturally biased and devoid of scientific merit.

But are the major differences between the United States and Europe in the acceptance of transgenic crops simply cultural artifacts? Are European concerns dictated solely by politics? And is Glickman right to imply genetically engineered crops raise no new scientific or safety issues that have not already been resolved?

Cultural Bias?

From a European perspective, opposition to genetically engineered crops is much more than a cultural artifact. Given the recent Ukrainian experience with Chernobyl and the fresh memories of misuse of German science in WWII, many new scientific manipulations of living things are intrinsically distrusted.

Glickman may be missing the deep historical precedent that underlies distrust of genetic science when he admonishes Europeans for their "Blind adherence to culture and history." Considering the mass annihilation of whole peoples based on faulty genetic science, it is understandable if Europeans view a "master race" of food crops skeptically.

Europeans are clearly looking at genetic technologies through a different cultural lens than are Americans. Part of this viewpoint is reflected in the phraseology chosen for genetically altered crops. While the United States government confuses the identity of genetically altered food crops by calling them "transgenic plants," Europeans call them "genetically modified organisms" (GMOs). This semantic distinction may be significant. The European community may find it easier to categorize genetically engineered crops and animals under the GMO rubric than are Americans who may be technically confused by the scientific jargon of "transgenic." At the core of the problem is the distinction between natural and artificial.

"Natural" versus Engineered

What is "natural" and what is artificial is of particular concern to those involved in the organic food movement. For many, this distinction is central to a world view focused on non-chemical means of production, limiting the amount of manipulation of the food chain. New regulations in the United States will permit limited use of the phrase "organic produce" to be applied to genetically modified food crops, on a case by case basis. But for many, the very fact that a "foreign" gene has been interposed in a natural genome is source of concern and disapproval.

Irrespective of their commitment to organic foods, large numbers of Europeans remain skeptical about genetically engineered crops. One survey, conducted by Professor Thomas Hoban at North Carolina State University in Raleigh, North Carolina, in 1995, showed that 49% of Swedes, 70% of Germans and 78% of Austrians would not buy a bioengineered food product. In contrast, an average of 70% of Japanese, American, and Canadian consumers would readily buy such products.[5] At a minimum, this disparity reveals a rift between Japanese and North American versus European attitudes towards genetically engineered crops.

At least part of the explanation for the European distrust of GMOs can be related to the growing environmental movement in Europe. The Green Party has encouraged this trend. In some countries like Germany, some commentators believe the Greens have built on a general antitechnology tendency. It is true that many in the Green Party believe certain technologies are furthering environmental depredations brought about by industry and nuclear energy, but its opposition to genetically engineered crops is more fundamental. The Greens and their supporters believe GMOs represent misuse of genetic power.

Other factors contributing to the antipathy towards genetic engineering have been identified by our correspondent, a social scientist in Sweden, Martin Frid. When we asked his view of the European position, he listed ten major reasons for European distrust of biotechnology (shown in Figure 4). Several of these observations, notably numbers 2–5, are especially salient for European concerns while others (numbers 8–10) characterize American attitudes as well.

From our perspective, at least three factors have fueled the growing malaise with genetic engineering in Europe. First, in Europe the distrust of things genetically engineered *is* clearly emotionally and culturally driven. This reality does not mitigate the depth or legiti-

Figure 4. Possible Reasons for European Mistrust of Biotechnology

1. A long history of natural, organic agricultural practices.
2. A growing recognition of recent failures of chemically based agriculture.
3. A distrust of scientists and their attempts to subdue nature through overreliance on chemicals like DDT.
4. A fear of monopoly control of European agriculture by non-European corporations.
5. A distrust of American regulatory agencies like the Food and Drug Administration which have given a clean bill of health to many products and chemicals that later proved harmful.
6. The loss of the small family farm.
7. Resistance from women who see American, chemically based, transgenic agriculture as a uniquely male operation.
8. Anger against the imposition of non-labeled crops and products by the United States on the European market.
9. Anger and distrust of the major companies who are developing biotechnology based agricultural commodities.
10. Secrecy and proprietary protections built into the American corporate way of doing business.

Source: Martin J. Frid

macy of popular concern. Ever since the Nazi atrocities, sectors of the European populace have questioned screening or health programs that carry the taint of genetic manipulation. Particularly in Germany, genetic control through in vitro fertilization, genetic engineering and screening, and cloning have all engendered deep misgivings and public anxiety about repeating the Nazi excesses.

In countries like Great Britain and The Netherlands, the public has rejected a previous genetically engineered product, recombinant bovine growth hormone (rBGH) stimulated milk, because of concerns about the safety of the resulting product. In this instance, many Americans shared this concern, while governmental agencies largely downplayed any difference between engineered and non-engineered milk products. The United States FDA asserted that no adverse health effects and no difference in growth hormone levels exist between milk taken from rBGH treated and untreated cows. Opponents pointed out, however, that the FDA failed to consider the secondary effects of forced feeding, higher antibiotic usage in some mastitis-affected cows, and the potentially higher levels of stress hormones in rBGH treated cattle. For others, like the founders of Ben & Jerry's Ice Cream in Vermont, rBGH treated milk is unacceptable simply because it threatens the small farm and is artificially made and engineered. Similar, "irrational" objections to transgenic food crops clearly play a role in the European population's reaction to transgenic crops. The basis for this opposition is instructive.

The European Perspective

European Union countries like Luxembourg, Austria, Germany and Italy have actively resisted the introduction of unlabeled, genetically engineered crops since 1992. The European Council of Ministers first drafted the so-called Novel Food Regulations on December 20, 1996, which specified the conditions for approval and labeling of genetically engineered foods. Three weeks later, these regulations were adopted by the European Parliament. On April 2, 1997, the European Commission voted 407 to 2 to ban the importation of unlabeled, genetically engineered American corn crops. As of July 31, 1997, European Union countries could import genetically engineered crops, but only if they were so labeled. The details of the directive call for a clear notification on the label that the product may contain or may consist of genetically modified organisms. For products consisting of mixtures of both genetically modified and non-modified organisms, the label must disclose the possibility that genetically modified organisms may be present.[6] The fine print of the present policy points up the political difficulties intrinsic in making such a determination: According to Article 8, which describes the scope of the regulation, a given foodstuff or crop is considered genetically modified and novel "if scientific assessment, based on appropriate analysis of existing data" demonstrates that it differs from "conventional food" in a significant way.[7]

According to one of our European contacts, French farmers have resisted novel genetically created crops in part because they simply dislike being told what to plant and how to plant it. This point is not lost to the detail men of major seed providers here and abroad who have been instructed to ensure that they put farmers "in charge" of new genetically engineered crops. But farmers will discover that in most instances they cannot keep or save any seed that is produced by their own crops. As we will discuss below, genetically engineered seed, especially the Roundup Ready™ technology, remains the property of the maker and not the user.

In Germany, in particular, the cultural bias against genetics has greatly reduced governmental and corporate investment in biotechnology, a position decried by Nobel laureate James Watson. In a talk to German politicians given in June 1997, Watson argued that Germany will have to get over its cultural malaise about genetics if it is to join the biotechnology age.[8] Watson asserted that European activist groups, especially those like the Green Party, have undercut genetic research. It is true that the Greens and their affiliates are concerned about domination by the United States-based multinational corporations, an issue brought out in sharp relief by the activities of companies like Monsanto which have extended their global reach to involve Europe, Brazil and Argentina. But, like us, the Greens do not have a single-minded view about a particular science, only concerns about its potential misuses.

Some Scientific Concerns

What merit do such concerns hold? As we will show, it is likely that genetic modifications made through transgene technology will not always be perfect or ideal. Some early genetic manipulations will of necessity disrupt the balance among the organism's genes needed for normal functioning. This has already proven true in animals where, for instance, inserting genes for human growth hormone in hogs led to misshapen, arthritis-ridden pigs. Similar, albeit less apparent, aberrations are possible in transgenic plants. In fact, many if not most genetically modified plants are the results of thousands of failed experiments. In some instances, transgenic plants were so disrupted by the gene insertion process that successful growth failed. Even where successful, few genetically modified organisms have been precisely modified. Few, if any, genetic agronomists can say where their gene ended up within a given plant genome. As a result, genetically modified plants may be much more of a black box than the pseudoscientific terminology of "transgenic" and implied genetic control connotes.

We will explore other reasons for questioning genetic technology below. Suffice it to say for now that we challenge the orthodox description of such plants as scientifically controlled wonders with stably introduced, balanced genomes.

Another potential problem is that the genes being manipulated are presumed to affect only single traits. But many plant genes produce a variety of effects (called "pleiotropy"), where changes in form and function result from a single gene insertion. Traditional breeding practices take such effects into consideration. In contrast to transgenic crops which are often marketed after only a single test plot is harvested, traditional crop breeding has been much more tedious—and careful. Only after careful breeding, often spanning three to four growing seasons, were novel varieties widely introduced. Norman Borlaug's dwarf wheat is a case in point.

A Pre-Biotech Case Study

Borlaug, a plant breeder at the Center for the Improvement of Maize and Wheat (CIMMYT) outside of Mexico City, bred a remarkable strain of wheat in the 1950s and 1960s through his efforts to increase cereal yields. He found that simple increases in soil nitrogen, while stimulating wheat growth, normally produced an unwieldy plant that was too tall for most combines to harvest and subject to wind damage or "lodging" where the growth-stimulated plant would fall over.

Through judicious breeding methods and introgression, Borlaug successfully introduced a "dwarfing gene" from a variety called Norin 10.

Originally isolated by indigenous farmers in Japan in 1873, this gene was introduced into Mexico by native farmers at the turn of the century. Wheat yields increased threefold. Borlaug simply transferred this existing trait (actually a group of closely linked genes) onto a modern wheat variety, giving it the ability to grow in harsher conditions with shorter growth times than before. According to a sign posted at CIMMYT, "100 million lives have been saved by Norin 10."

While this latter statement may be slightly exaggerated, it does reflect the dramatic increase in yields brought about through judicious breeding in the early years of the Green Revolution. The key question about genetically engineered crops is whether or not similar advances could prove attainable through the systematic introduction of single genes. While some, like USDA's Glickman, tout transgenic crops as promising a second Green Revolution, those who control the technology appear less interested in meeting world food needs than in obtaining short term goals. More to the point—would present day genetic engineering science be up to the task?

Literally thousands of transgenic plants have already been sown and millions of gene-altered seeds produced. The genes being chosen for engineering are in the main quite different from Borlaug's Norin 10. Instead of blocks of genes that will increase yield or improve efficiency, geneticists usually inject only one that confers an aesthetic change or makes the plant tolerant to a proprietary chemical. Any increase in yield or nutritional value is usually incidental to a more readily achieved economic advantage. In fact, as we will see for Roundup Ready™ crops, some studies indicate *reduced* yields for some strains of genetically engineered crops when compared to the non-genetically engineered crop lines. This point is central to addressing the claim that genetically engineered crops will solve the world's food problems.

World Views Contrasted

Our position is simple: We are skeptical about pronouncements from government officials and optimistic commodity experts that tout genetically modified crops as a panacea for the world's food shortage and a solution to overreliance on pesticides. We see transgenic crops more as an extension of corporate dominance, one centered on short-term gains for shareholders. Of course, the corporate view (to be documented below) is a more sanguine one. While the ultimate truth behind these viewpoints may never be elucidated, one thing is certain: These two world views, one skeptical about our wisdom in modifying life forms, the other highly optimistic about the extension of our technological power, are headed for a collision. Many people, particularly those affiliated with the environmental movement, are profoundly skeptical that we will "get it right" this time. The time honored tradition of subduing nature and remolding it to fit human needs has left thousands of acres in the Sahel region of Africa and our own Southwest region desertified and useless. The initial success and then horrible debacle from overuse of DDT is another example. Even seeming miracles like antibiotics have been despoiled from misuse or overreliance. Some religious figures declare the belief that nature is intended for "Man's [sic] use" engenders a patriarchy of control that misconstrues the purpose of life on earth. Others decry the attitude that casts all living things as intrinsically designed for our exclusive use.

The truth, of course, is that the natural world was never intended or designed for our exclusive use, nor shaped to fit our nutritional needs. Such a realization is belied by our long history of domesticating animals and food crops, a grand experiment that many will cite favorably when considering genetic engineering. But, even as we domesticated plants and animals to reshape them into useful forms, we discovered that many food crops and animals harbored toxins and pathogens that undermined their intrinsic domestic value. The intensification of production that came with the mass growth of livestock in particular has proven ill-advised. Antibiotic resistant bacteria from feed lots, hormone-laced (DES) beef in the 1960s, and bovine spongiform encephalopathy (BSE) today, are all the result of mass production pushed ahead of good science and epidemiology. Mad cow disease is a sobering reminder of the pitfalls of our overreliance on human control to domesticate animals en masse.

Our concern is that the new genetic revolution in agriculture threatens to be similarly disruptive. If pressed ahead at the present accelerating pace, transgenic agriculture may further disrupt our reliance on traditional farming techniques with uncertain ecological impacts. At the core of the problem is the folly of massively substituting human engineered crops for those that were previously selected by natural forces. We do not reject all manipulations as being intrinsically wrong because they alter nature. But we question the blind assurances that the new wave of genetically engineered crops promise unlimited returns at virtually no environmental cost.

Scale and Scope

The very scope and breadth of the revolution to genetically control crops portends radical changes. We are concerned with the vast range of implications of such widely used genetic engineering of our food crops. These concerns span from possible impacts on our health, to wholesale cultural and ecosystem disruption. Are the risks that are being taken commensurate with the benefits we may expect? Will these risks and benefits be distributed equitably? Are we alert to the possibility that as we expand our crop-conversion to novel bioengineered types we may reach a threshold of vulnerability? For instance, many countries like India are right at the balance point between survival and famine. A loss of 15% of the rice crop could endanger as much as 30% of the population. The answers to these problems require answers to a set of prior questions:

1. Are genetically engineered crops simply extensions of a long history of domestication of wild strains of plants, or do they represent a qualitative departure in the way we grow our food?
2. Do genetically engineered crops signal a true revolution in our stewardship of the planet?
3. Will bioengineering offer needed crop productivity at a period of continued population growth?

In the end we must make any decision to proceed in a state of enlightenment rather than in a state of desperation. It is with this view in mind that this book is written.

Ultimately, we are challenging the premise that biotechnology will make things better in the long run, not just in terms of the immediate increases in yields or reduced pesticide reliance promised by the innovators, but in terms of the overall sustainability of agriculture as a whole. To fully appreciate the magnitude of the potential problems generated by these new technologies, we clearly need to know the scope of the enterprise. We need to understand how innate cultural resistance to new crops is being overcome by public relations efforts. We need to understand the motives and practices of our major corporations. And we need to be able to anticipate the secondary and often unforeseen consequences of their widespread "success."

In doing this research, we have relied on both corporate documents and major scientific publications. Governmental agencies and their underlying regulations have also been consulted. Hopefully, we have synthesized this information in a form readers will find useful in formulating their own opinions.

---

[1] Nordlee, J. A., Taylor, Townsend, J. A., Thomas, L. A., Bush, R. K. "Identification of a Brazil-Nut Allergen in Transgenic Soybeans," *The New England Journal of Medicine,* 334: 688–692, 1996.

[2] For 1997, this estimate includes 9 million acres in Roundup Ready™ soybeans; 6 million in Yieldgard® Corn; 6 million acres of Bollgard® Cotton; and 2 million in Roundup Ready™ Cotton; an estimated 2–4 million in other genetically engineered crops are known to be planted.

[3] Keith Thompson, Vice President General Manager of Hartz™ Seeds, personal interview, June 20, 1997.

[4] Remarks of Dan Glickman, Secretary of the United States Department of Agriculture, 20 June, 1997.

[5] Hoban, T., "Public Perception and Communication of Risk," Raleigh, NC: NC Cooperative Extension Service, 1991.

[6] Amended to Annex III of Directive 90/220/EEC of the European Commission of the European Union.

[7] See Emma Johnson, "European politicos pass food regulations lacking meat," *Nature Biotechnology* 15: 21, 1997.

[8] James Watson, quoted in R. Koenig, "Watson urges 'Put Hitler behind us,'" *Science* 276: 892, 1997.

1

The Argument for Biotechnology

Population researchers estimate that by the year 2025 there will be 8.25 billion people on the planet. Agricultural based companies have cited these statistics as part of their rationale for devising seeds for superior yield as well as for better pest/disease resistance, a point we discuss in Chapter 7.

The United States has become a global leader in the development of bioengineered food crops designed to feed the burgeoning population. United States-based companies such as Monsanto, AgrEvo and Dupont are all developing technologies that will be resistant to their chemical products. The current United States government's regulators are silently supporting this engineering movement, in marked contrast to European authorities.

Chemical companies that make agricultural herbicides know that as transgenic plantings increase, so will the need for their herbicides.[1] Ideally, herbicide tolerant crops require fewer herbicide applications because the targeted crop is programmed to "accept" a single chemical product. Many large companies who are now involved in both the planning and chemical aspects of agriculture capitalize on this fact. For example, Monsanto Company, a major chemical manufacturer, has recently become active in developing engineered technologies for crop seed. They have also been very busy in the past few years buying large portions of seed companies in order to market their new technologies. Since 1992, Monsanto has acquired whole or part ownership of six major seed producers in the United States.

Another argument used in favor of biotechnology is that the promise of greater yields preserves precious habitats that would otherwise have to be destroyed by increased farming. In fact, two-fifths of the world's current food production already comes from the 17% of land that is irrigated.[2] High precision agriculture could reduce land needs dramatically *without* turning to transgenics.

The Scope of the Problem

Without a doubt, biotechnology influenced agriculture—what is known in the industry as "agbiotech" is big business. For some entrepreneurs, it is a golden opportunity to amass wealth and garner greater control over farming than ever before. From the viewpoint of the growing numbers of agbiotech consultants, genetically engineered crop systems represent an unheralded investment opportunity. In the words of the president of a major Canadian agbiotech consulting company, "The scope of biotechnology as it is used in agriculture will touch every segment of society in every country in the world and surpass anything we have known in the production of food."[3] The aggregate dollar value of all agbiotech inventions is likely to exceed $46 billion by the year 2000, according to a consultant at Ernst & Young of San Francisco.[4] Left unsaid is the likelihood that any increase in intrinsic value is due to enhanced production of animal feed crops like soybeans and corn and not to an increase in foods directly consumed by people.

The countervailing view, that agbiotech's major promise is to feed the masses, is represented by statements made by Dan Glickman, the Secretary of the USDA, in remarks given to the international Grains Council in London, England, on June 19, 1997. Glickman claimed, "There is no way to feed a hungry world, or an economically growing world, without embracing the future. We must make this choice [for biotechnology]. We have an obligation to our world, and it cannot wait. If we leave these decisions to future generations, it may very well be too late."[5]

But is this urgent view of agbiotech as a panacea accurate? Will biotechnology afford an unprecedented vehicle for advancement?

Or will it fail to meet these early glowing evaluations? Part of the answer hinges on the applications of biotechnological advances in agriculture. To date, biotechnology has been applied to a number of innovations that have made agricultural products more "consumer friendly," but few have genuinely increased productivity. Many of the genetically engineered feats are inventions that change the ripening or shipping characteristics of plants, such as making aesthetic changes in tomatoes or cantaloupes that improve their consumer acceptance. In 1994, marking the first instance of a transgenic product going on the market shelves, Calgene attempted to commercialize the Flavr-Savr® tomato that had delayed ripening characteristics permitting the simultaneous harvesting of ripe, firm tomatoes. But poor consumer acceptance led to failure. The same year DNAP Holding Corporation of Oakland, California, also developed a product, the Endless Summer® tomato, promising a better flavored, firmer fruit. It, too, did not meet consumer's expectations, perhaps because of poor marketing. As a result, both the Flavr-Savr® and Endless Summer® tomatoes were removed from the market "for further development."

Still other products are in the offing that promise greater consumer acceptance. Examples include bell peppers with longer shelf life, peas with higher sugar content, and controlled ripening Roma tomatoes. By controlling the amount of ethylene released by ripening fruit, biotech firms also hope to dampen the rate of ripening, extending a commodity's shelf-life. This was in fact the technology used to develop cantaloupes with longer commercial appeal.

On the sidelines, less promoted but more essential developments are also underway. Several companies are trying to develop corn and wheat lines to be more insect and drought resistant. These developments hold potential advantages for agriculture, and if successful, promise much higher yields under a variety of conditions. At the other extreme, chemical company scientists are devising means of giving plants *both* a resistance gene and chemical tolerance to herbicides like Roundup® and bromoxynil based Buctril®. In this manner, agronomists hope to be able to simultaneously control pest and weed problems. The targeted product lines will include corn, wheat, cotton and rice. As we will detail below, these approaches have certain intrinsic weaknesses as well as benefits.

Evolutionary Perspectives

By moving from the occasional tinkering of molecular geneticists to wholesale manufacture on a grand scale, we may be courting unanticipated evolutionary transformation. Where an occasional genetic introduction to improve insect resistance was once a limited change, we now have entire fields of cotton and corn growing where the entire cellular makeup of each plant cell expresses insect resistance toxins. The resultant overkill has already duplicated the scourge of antibiotic resistant genes in bacteria—too much of a good thing breeds resistance—almost overnight. In the instance of the first field trials of insect resistant cotton in Texas, the fields were virtually overrun with bollworms by the end of the growing season for as yet unknown reasons. Now farmers will be required to leave a major portion of their fields (up to 20%) planted with non-genetically engineered varieties to give nonresistant insects a refuge from total annihilation and resulting selection pressures, a point discussed below.

A similar pattern of domination exists in the pesticide sector. One of the ways scientists have genetically engineered crops is to make them impervious to the weed-killing chemicals that might be sprayed on a field. This genetic ploy works to allow the maximum use of broad spectrum herbicides with impunity, sparing the crop while killing the weeds. As we have shown, it is no accident that most of the seeds grown by Monsanto subsidiaries are engineered to be resistant to its own herbicide, known in the trade as Roundup®; not coincidentally, the genetically engineered crops are called Roundup Ready™.

As we will show, in just the last two years Monsanto has come to dominate the world's herbicide production, making 26 million pounds of Roundup® in 1996 alone. While this herbicide is considered relatively safe for birds and animals, its toxicity for humans is still incompletely understood. Nonetheless, to permit Roundup® sprayed crops to be used for livestock fodder, USEPA has allowed an increase in glyphosate residues (the active ingredient in Roundup®) tolerances to go from 6 to 20 parts per million for raw soybeans and from 100 to 200 parts per million for soybean hay and 100 parts per million for their hulls. The possible ecological consequences of this move will be explored in Chapter 3.

A Corporate Perspective

At first glance, changing the genes in crop plants to make them more pest resistant or herbicide tolerant appears to be logical. Some plants, like those in the chrysanthemum family, naturally contain high concentrations of insect-repelling chemicals known as pyrethrins. Imbuing a valuable commodity like corn with Bt genes for a toxin noxious to corn borers is presented by its developers as an extension of this natural stratagem because Bt expresses naturally occurring toxoids that are poisonous to insect pests. Similarly, providing soybeans with herbicide resistance genes gives the farmer immensely greater latitude in spraying his fields with weed-destroying pesticides and reduces the need for multiple sprays.

Given these realities, many industry executives are smarting over what they believe to be unjustified criticism. Corporate representatives with whom we have spoken criticize what they see as an arbitrary disregard of the true benefits of these engineered crops. From the industry point of view, bioengineered crops are both reasonable and economically sound. For them, the intrusion of genetic material into feed crops or cereal grains is nothing more than a logical extension of the process of plant domestication.

An Environmental Perspective

We don't think so. Placing new genetic characteristics in plants once required years, if not decades, of selective breeding and culling of unwanted varieties. Today, it takes a matter of months. Unlike Norman Borlaug's laboriously derived strains of dwarf wheat or rice which spawned the agricultural revolution of the 1960s, the new genetically constructed crops are designed for a single technological advantage, such as herbicide resistance, a trait which may or may not ultimately translate to improved yield or quality. In the rush to market, cultivars with unintended properties may emerge.

From an environmental perspective, herbicide resistance may also carry a unique downside. Two-thirds of the newly marketed bioengineered crops are being created to be herbicide tolerant, meaning that otherwise toxic amounts of a given herbicide can be oversprayed on the protected crop to kill surrounding weeds while leaving the commodity virtually unscathed. In some cases, herbicide residues will rise in the edible portion of the crop. Such an eventuality virtually assures that consumers will be exposed to more pesticide in the genetically engineered product than in the prior one. The argument of industry representatives that these newly engineered herbicide resistant varieties would actually reduce pesticide exposure remains unproven, especially in the light of the apparent necessity for increased tolerances. (We will explore this point in detail in Chapter 5.) For now, we note the incentives for overuse of the crop protected pesticide are intense since clearing weedy competitors virtually guarantees greater yield.

Consumer Risks

Bioengineered crops may also pose hidden risks to the consumer. In the case of herbicide tolerant crops, the foods we want will not only likely have higher pesticide residues of the "resistant" chemical, but the food itself will contain the newly introduced gene and its products. Unless engineers selectively direct gene expression away from edible plant parts, crops created to be resistant to *Bacillus thuringiensis (Bt)* will contain the *Bt* toxin which consumers will ingest along with their food. This means we may be eating *Bt* toxin *and* the resistance proteins for glyphosate and/or a more toxic herbicide, bromoxynil, along with the nutritional products in the seed itself. Such a situation not only poses a largely ignored—and we feel major—problem to us, but also for our native intestinal flora and fauna that will be exposed to edible oil or seed derived from crops, especially cotton and soybeans.[6]

But the government has resisted labeling bioengineered commodities as being derived from genetically engineered plants. As long as the engineered products remain largely anonymous, we believe the public is at the mercy of corporate ethics. And as long as the bottom line remains preeminent, the problems generated by genetically engineered monocultures and monopoly will endure. Most agricultural scientists recognize that for maximal stability and long-term yields, we need not more genetic control and homogeneity, but continued variety and diversity. Local control to meet local microclimatic conditions virtually dictates genetic diversity and freedom of choice. But with genetically engineered seed projected to be wholly introduced into our crops, an evolutionary disaster from lost genetic diversity, or other unforeseen consequences to soil or plant ecology may loom just around the corner.

We believe the coming genetic revolution can be an enlightened one only if the public understands its implications from its inception.

---

[1] Labor Department and Federal Reserve Board Calculations. *Chemical and Engineering News*. 31 March, 1997.

[2] Editorial. *Lancet* 348: 1109, 1996.

[3] Meg Claxton, quoted in *Genetic Engineering News,* 15 June 1997, p. 6.

[4] Cited in Gail Dutton, "Agbiotech companies' technologies lead the way toward a carbohydrate-based economy," *Genetic Engineering News* 17: 1,6,39, 15 June 1997.

[5] Dan Glickman, Remarks of USDA, International Grains Council, London, England, 19 June 1997.

[6] Technically, we already ingest some products in corn, grasses, or other monocotyledenous plants that make them resistant to herbicides like 2,4-D that attack dicotyledenous plants like dandelions.

2

Profiling Genetic Engineering

In 1998, no one doubts genetic engineering is a revolutionary technology which permits unprecedented human control of the innermost workings of living organisms. While in its infancy, this new technology opens a wide spectrum of potential manipulations of life forms by intentionally modifying their basic genetic instructions. If performed at the level of seed production, the resulting genetic changes will be permanently encoded in the germ plasm of all descendants of the manipulated plant. It is this permanent effect which offers the greatest promise for this new technology and its greatest peril. Any mistakes made early in the genetic engineering game may reverberate for decades.

Practitioners of genetic engineering technology insert genes with known characteristics and/or products into a strain of plant or animal previously lacking the desired trait. The newly composed DNA of the host is called "chimeric" DNA, from the Greek *chimera* of mythology, a creature with the head of a lion, body of a goat, and tail of a serpent. The term "engineering" is used for the construction of artificially recombined molecules because they produce structural combinations of DNA by biochemical means.

In the case of the genetically engineered food crops, representative introduced genes include those that make new or higher quantities of essential amino acids, produce certain enzymes used in photosynthesis, or make proteins that confer resistance to pests. The relative ease with which genetic engineers get plants to take up and use exotic genes underscores a fundamental truth: all genes, whether in plants, animals, prokaryotes or eukaryotes, use the same language. In this sense, the DNA sequences are the Esperanto of biology, a kind of universal tongue that is recognized by all organisms on Earth.

The key to getting this language to be expressed and read in a new organism is getting the appropriate DNA sequences into the recipient organism's own genetic stock—its chromosomes—along with appropriate "start" and "stop" instructions to permit the transferred gene to be "read" properly by the recipient's molecular machinery. As discussed below, this feat is not so much a technological marvel as it is opportunistic. New genes are piggybacked onto existing genomes—the full genetic makeup of an organism—by brute force. The few plants in which such genes "take" are the ones chosen for propagation. Often the insertion device is simply a preexisting well-worn agricultural bacterial or viral pest that already "knows" how to insert genes into plants. More recently, genetic engineers have used a literal "shotgun" approach, firing microbullets coated with DNA into plant target cells.

Origins

Each of these technologies has its origins in the achievements of a few genetic pioneers. In 1927, a medical officer named Frederick Griffith, from the British Ministry of Health, made a revolutionary discovery: He isolated two different kinds of pneumonia-causing bacteria from one of his patients and then inoculated mice with both resulting strains. One of the strains produced a fulminating infection while the other was virtually innocuous. Had he left his observations there, modern genetics might have waited another decade. But in one experiment, Griffith combined an extract of the killed, virulent bacteria from the first strain and *mixed it* with living organisms of the benign strain. To his astonishment, the benign strain acquired the host-killing attributes of its dead progenitor! Something in the cell-free homogenate of the virulent bacteria had conferred the lethal properties of the virulent strain onto otherwise safe bacteria, transforming them into killers.

This was the first known instance of genetic engineering. Fifteen years later Oswald Avery and two of his coworkers at the Rockefeller Institute in New York ferreted out what this "transforming principle" was made of. Avery and his colleagues discovered that it was a nucleic acid—and not the proteins previously suspected. Avery was the first to show nucleic acids were responsible for carrying hereditary information from one cell to another. This research led to the identification of deoxyribonucleic acid, or DNA, as the genetic material and, serendipitously, to the discovery that DNA could "work" even after it was taken out of a live environment and put back into living cells.

In 1953, Watson and Crick predicted that Griffith and Avery's chemical would be made of a long chain of bases arrayed along a phosphate backbone in the form of a double helix. In short order, they found the previously reported equality in numbers of certain base pairs (adenine and thymine: guanine and cytosine) reflected their cross-pairing, making the rungs that linked the helical chains. The sequence of bases along this helix spell out the genetic code, a code made of four "letters" or bases: *A* for adenine; *T* for thymine, *G* for guanine; and *C* for cytosine. In the language of DNA, each sequence of three bases "spells" an amino acid.

Among the other key discoveries which fueled the evolution of rapid genetic engineering was the fact that DNA base sequences could be isolated and synthesized in long chains. Another was that certain viruses—both the phages that infect bacteria, and plant and animal viruses that infect living cells of higher organisms—readily inject their own DNA sequences into a host. Some of these DNA chains became stably integrated into the host's genome, setting the stage for the usurpation of cellular machinery to make more virus particles, and creating the precedent for artificial genetic engineering. As with these natural infections, genetic engineers soon learned that success requires that the novel genes inserted into these foreign genomes replicate along with the host's chromosomes.

To artificially insert a set of novel genes into a host, it is necessary to have a controlled sequence of DNA, a portion of DNA called the replicon, which permits the survival of the introduced gene in its new environment, and a suitable introduction system, known as a vector. When all of these elements work in harmony, new genetic information is introduced into a cell's DNA, the new DNA is encouraged to transcribe and produce its protein product, and the introduced gene is stably integrated into descendants of the first cells and ultimately the plant itself. When all of this happens successfully, the original plant is transformed into a "genetically modified organism," or is said to be "genetically engineered."

The propagation of genetically engineered plants is usually achieved by a process of asexual tissue culture growth whereby single cells are coaxed into forming a multi-organed embryo and ultimately a miniature plantlet in a solid agar medium. Thereafter, the adult plant is transplanted and allowed to mature sexually, where, ideally, it passes on its modified genome. This occurs by the assortment of the necessary gene through meiosis into the gametes (pollen or egg cells) used by the plant for replication. Because the newly introduced gene is commonly only on one DNA strand, during meiosis the gene-lacking strand is also produced. Hence only half the pollen and ovules will contain the engineered product, making crosses necessary to make the gene homozygous so that each cell contains a double copy of the gene. This can be done by self-pollination or taking cells or cuttings from the original plant.

Specific Tools

One bacterium was particularly suitable as a vector to put novel genes into plants. Known as *Agrobacterium tumefaciens*, it does so because of special genes it uses to make a tumor known as the crown gall, a knobby growth that afflicts pines and other trees in coastal areas of Maine and elsewhere. Because scientists had learned how to insert novel genetic material into this and other bacteria, they were presented with a wonderful opportunity: Why not program the *Agrobacterium* to carry human-selected genes into its normal plant hosts? This possibility was first realized in the 1980s but was initially limited to the host range of *Agrobacterium,* namely, dicotyledonous plants. Monocots, like the grasses and cereal plants (the Graminae) that include wheat, rye and corn, were outside the infective range of this bacterium and hence initially outside the type of plants that could be engineered. Since about 1987 it has proven possible to infect monocots, especially the yam, with *Agrobacterium*, opening a new range of hosts for genetic engineering via this vector. Full success of *Agrobacterium*—mediated transformation in the cereal plants is still unrealized.

Even when specific gene sequences could be introduced into plants, they were limited to tiny plant embryos. But because plants replicate agonizingly slowly compared to bacteria and require special culturing technologies before the tiny seedling can become a multicellular plant, it would often be weeks or months before the success of any given experiment was assured. Then plants with the targeted genes had to be propagated and selected for the presence of the desired characteristic. One way to accelerate this process is to link the desired gene to one for antibiotic resistance. By screening for those plants which carry the resistance marker gene, researchers can readily isolate the genetically altered plants they want.

A similar accelerated selection process can be used to isolate engineered plants with genes inserted by other means. One novel method is to literally shoot microparticles of DNA-coated gold, tungsten or other inert materials directly into plant cells. Initially, a live gunpowder charge was used to propel the tiny (four micrometers) particles, although recently both electric charges and air have been used successfully. This improbable microscopic bombardment with DNA covered particles (known as biolistics) has proven remarkably successful. By chance, some of the plant cells and organelles that are hit by this friendly fire will pick up the foreign DNA and, in as yet unknown ways, incorporate the new genes into their own nucleus. In a few instances, this "shotgun wedding" of new and native DNA can lead to a stable marriage, as the engineered strands become integrated into the plant nuclear DNA.

But this scattershot approach has its downside: No one knows where the novel DNA has been spliced into the plant's own array of genes. Only by blind luck will a few plants survive, and those that do will have this new DNA in different regions of their genome. Predictably, some of this new DNA may unbalance or disrupt the normal functioning of the resident genes, much as an unwanted guest can disturb a stable household. Should this happen, it may be several generations before the resulting disruption is fully realized. In the interim, the affected plants may show reduced vigor or other debility resulting from an imbalanced genome.

Many transgenic seed companies must therefore gamble that their first few "takes" represent plants with good internal balance and survival characteristics. With crossed fingers, genetic engineers hope their new contribution to the genome will produce no or only minor disturbances in the expression of native genes vital for normal functioning. Because of the press of time, few if any of the novel plants are tested for more than one generation to determine if the gene is acting in typical Mendelian fashion before seed production goes into full swing, although several plots are often grown simultaneously. Hence, the newly engineered plants are carried by whatever genetic balance existed in their prior genome. However, if, as with cotton plants, each new generation of plants is generated by new hybrid crosses (F_1; F_2; F_3; etc.), some genetic reassortment is inevitable, putting each succeeding generation at risk for destabilizing effects from repositioning of the novel gene sequence. (As we will show, something like this may explain the demise of a portion of the 1997 Roundup Ready™ cotton crop.)

The end result is transgenic plants or organisms that have been genetically bred to contain a novel gene that would not otherwise be present through natural methods of reproduction—and an uncertain commingling of the new with the old.

Examples of Transgenic Plants

Through the use of biolistics, dramatic alterations in plant morphology, biochemistry and chemical tolerance have been achieved. Among the more critical developments has been the successful introduction of genes that create tolerance for herbicides. Examples include the *bar* gene which provides a selectable marker for transgenic cereals. The *bar* gene codes for an enzyme which breaks down several related herbicides, including phosphinotricin or PPT and related congeners known as glufosinate and bialophos. Without the parent herbicide, *bar* inhibits the enzyme glutamine synthetase, preventing the plant from discharging ammonia. The resulting buildup of ammonia in plant tissues causes cell death. Those successfully engineered plants with the desired gene survive after herbicide treatment, denoting the presence of both the engineered and marker genes.

Another technique used in genetic engineering is a special tagging process by which transferred DNA is randomly inserted into the plant genome, resulting in inactivation of native plant genes. Such inactivation may actually have beneficial effects since taking out a gene that produces an undesirable characteristic can often be as useful as putting in a gene that offers a beneficial one.

Ultimately, it is this same random insertion process that has proven to be the undoing of some of the early genetically engineered plants. As we saw with biolistics, without knowing where the new gene is, researchers are literally shooting in the dark. Damage and disease may become evident only after several generations when such gene insertions create subtle plant morphological or biochemical changes which put the plant in harm's way for injury by a novel pathogen or environmental stressor.

A final type of transforming agent is a plant virus. Like viruses of animal cells, plant viruses could in theory be used to stably integrate their genes—and any attached carry-ons—into the plant genome in specified locations. Such engineering has been proposed to allow the massive infection of plant cells (even with their death) for the successful harvest of an engineered gene sequence product from the dead biomass. Another option is to produce "more-or-less" healthy plants in which a foreign protein is expressed which might make the plant more palatable, nutritious or otherwise desirable. It is also conceivable that virally engineered plants could be made that produce a vital protein for human consumption: The bioengineered banana with bacterial proteins that can be taken as a vaccine (hepatitis B) is an example. So is the plant which carries oral "toleragenic" proteins into the body to combat the autoimmune form of diabetes. Examples for viral transfection of plant cells are cauliflower virus, brome mosaic virus and the intensively studied tobacco mosaic virus, although their genomes and size are too small to allow the large-scale gene introductions possible with *Agrobacterium.*

History and Perspective

The use of some form of genetic control *per se* to alter the properties of crop plants is not a new phenomenon. Since humans began agricultural practices, primitive peoples have been selecting seeds that would produce crop variants that would meet the special needs of their cultural communities. Fourteen thousand years ago root foods crops like yams were domesticated in the Old World. Ten thousand years ago the first squash plants were grown in what is now central Mexico. Since then, human innovation has been limited to permitting the movement of new genes from the wild into carefully selected stocks (introgression) through judicious use of outbreeding and selectively fertilizing and hybridizing the most desirable varieties to assure a crop with desirable properties.

Early examples of breeding success dating back some 10–12,000 years include sweet potatoes and cassava in Africa and potatoes and maize in the New World. In each example, indigenous peoples have worked assiduously to bring crops into harmony with specific microclimates and ecosystems.

Plant Heredity

Beginning about the turn of the 20th century, the hereditary basis for critical traits like color, texture and seed size in peas were discovered. In 1858, an obscure Bavarian monk named Gregor Mendel first reported that these traits could be inherited in relatively simple combinations. Mendel's work, which led to the recognition of dominant and recessive traits and the beginnings of modern genetics, was only recognized some 50 years after his original experiments were completed.

In 1913, William Bateson coined the term "genes" to describe the physical units that carried traits. Later, in the 1920s, Luther Burbank (1849–1926) successfully created hybrids that carried traits of related plants into a common stock. By assuring that such plants were kept "pure" through genetic control, breeders were able to patent common varieties and sell their cuttings. In this, breeders established the principle of privatization and set the precedent for commercialization of seed stocks. This tradition has been extended to permit the patenting of plant varieties and the control of genetically engineered seed stocks. We note the conversion of conventional crops to engineered crops has happened due to decisions in the board rooms of companies, resulting in crops with greater earning power and not necessarily greater nutritional value.

Human Intervention

The alteration of nature by human intervention is of course neither new nor necessarily dangerous per se. Charles Darwin, in his seminal work, *Plants and Animals under Domestication,* emphasized how long intentional selection was used to improve food crops. He observed, "We now know that man [sic] was sufficiently civilized to cultivate the ground at an immensely remote period; so that what might have been improved long ago up to that standard of excellence which was possible under the then existing state of agriculture."[1] Studious selection of seed corn was recommended as early as Roman times. Even Virgil noted that without such intensive and continuous selection, degeneration of a crop (that is, reversion to less well adapted types) was inevitable. As he wrote, "I've seen the largest seeds, tho' view'd with care, Degenerate, until th' industrious hand Did yearly cull the largest."[2]

Thus, the necessity of continuous human intervention to maintain the quality of selected crops was recognized in antiquity. This reality contrasts graphically with the modern sociological observations, documented previously, that many members of the public carry a high degree of anxiety and concern about technological interventions and direct manipulation of nature to maintain or improve contemporary crops. Given the commonality of this practice, a wholesale rejection of agricultural control appears irrational and unjustified. Just what is it then that makes direct genetic manipulation questionable?

As we have shown, clearly some of the antipathy towards genetic control *is* psychologically and culturally driven. Many people in the U.S. and Europe are highly skeptical about genetic intervention of any kind. Others are especially wary about human genetic engineering. A 1995 Field Poll of public attitudes about genetics conducted at the behest of the March of Dimes revealed that a majority of American adults are profoundly ambivalent about using genetic technologies. Many are reluctant to use genetics to find out more about themselves, and others are highly concerned about the potential for abuse should genetic engineering become a reality. From our perspective, the roots of public anxiety about crop engineering are sown in two issues: monopoly control and secrecy.

Major Actors

Today, a few major chemical corporations vie to control genetically engineered varieties of wheat, corn, potatoes and cotton. Major chemical companies like Dow, DuPont and Monsanto have jumped into the "life sciences" industry and are also aggressively developing genetically engineered products. These industrial scientists who once developed agricultural chemicals are now expanding their markets by creating crops that will become dependent on the same chemicals. As we showed, Monsanto Company is creating crop plants to be resistant to their high selling Roundup® herbicide. Their Life Sciences division has been largely funded and developed from sales of Roundup® herbicide. Large numbers of new varieties of corn and soybeans are being fabricated with genes that determine traits for herbicide tolerance and, to a lesser extent, insect and other pest resistance. In theory, such traits could be used to enhance commodity production because they produce crops that require less reliance on pesticides and have greater intrinsic disease resistance. An example is the Freedom II Squash which was spliced with viral genes to increase disease resistance and increase yields. As we emphasized, the vast majority of new transgenic crops are engineered for herbicide tolerance rather than for any intrinsic improvement in crop food quality or pest resistance. In spite of this bias towards chemicalization (a fact which we document below), our government supports biotechnology as an unquestioned good.

Potential Problems

In a talk at Tufts University Medical School in early 1997, Carol Browner, the chief administrator of the Environmental Protection Agency (EPA), has applauded the virtues of biotechnology and touted this science as offering the best hope for enhanced productivity and decreased reliance on environmentally damaging chemicals.[3] Is this unbridled optimism justified? The history of technological invention is riddled with examples of premature commercialization with belated testing and evaluation. Many potentially hidden pitfalls plague the intense race to establish corporate dominance in any new field. For instance, the development of silicone breast implants, the Dalkon shield and the Björk-Shelley heart valve all proceeded with minimal or no long-term safety testing. We may chart a similar error in the programming of crops to express a bacterial toxin from a species (*Bacillus thuringiensis*). This bacterium previously occupied a relatively small niche among soil microorganisms. The proliferation of its genes in plants will almost certainly generate many new evolutionary problems for organisms including insects, both beneficial and harmful. We will follow this story in Chapter 3.

The rationale for focusing concern on such genetically engineered crops and not on agricultural practices more widely, hinges on the fundamental difference between bioengineered and conventional crops. In genetically engineered food crops a single gene is introduced into the genome of a plant to produce desirable characteristics. In traditional breeding methodologies whole blocks of genes are moved by allowing crossovers or meiosis to separate and perpetuate genes conferring useful properties. Both new and old methods can change the genetic composition of field crops, but only the latter is a proven means of ensuring some semblance of uniformity from generation to generation, a point underscored by Norman Borlang's successes.

With genetically engineered crops, investors are gambling that the *one* chosen gene and its genetically uniform hosts will be kept in commerce for at least a decade. But the natural cycle for hybrid varieties is no longer than seven years before nature catches up to them and new pathogens or pests make continued planting infeasible. The plain truth is that the newest genetic strategies portend massive changes in the way we produce our basic commodities without any assurance of permanent or even lasting benefit. In theory, *Bacillus thuringiensis (Bt)* derived resistant crops are desirable because they carry an innate pest resistance conferred by a toxin ensconced in each of their cells. Such an introduction usually makes the plant insect resistant throughout. But the very prevalence of this "advance" may be its undoing. *Bt* engineering assures *Bt* toxoid will be found throughout plant tissues. As we discussed, this distribution virtually assures mass exposure of any pest or pathogen to low-level toxin over their life cycle, greatly accelerating the emergence of resistant pests. The *Bt* strategy works best *if only a few crop plants* are insect resistant. But when an entire field of cotton is sowed with *Bt* resistant genes, the net effect can be counterproductive: over time, only resistant pests will survive to reproduce.

Such a scenario is more than hypothetical as we will show in Chapter 3. For now, it is sufficient to observe that wholesale production of genetically engineered crops is proceeding with hardly a backward look at their long-range evolutionary fate or consequences. Literally dozens of transgenic commodities are being developed and released on millions of acres in the United States alone. Some representative products are listed below:

Examples of Products in the Pipeline

1. Liberty Link™ has been developed by AgrEvo, formed in 1994 with the merger of Hoechst-Roussel Agri-Vet Company and NOR-AM. Together these companies make the fourth largest agricultural chemical company in the world. AgrEvo means "Agricultural Evolution in Action."

   This genetic technology is designed to be specifically tolerant to DuPont's Liberty® herbicide. The active ingredient is glufosinate-ammonium (glufosinate ammonium:butanoic acid, 2-amino-4-(hydroxymethylphosphinyl)-,monoammonium salt). As described previously, this herbicide is one of the two which act by preventing the detoxification of ammonia, leading to a buildup of ammonium which disrupts photosynthesis, causes cell death and ultimately the death of the entire plant. Liberty Link™ transgenic plants containing the additional gene from a strain of bacteria *Streptomyces viridochromogenes* are sufficiently resistant to this otherwise powerful herbicide to allow the "over the top" application of Liberty® herbicide during the growing season for weed control with minimal or no impact on crop yield.
2. BXN® Cotton technology was developed by Calgene, Inc. (now owned by Monsanto Company) in conjunction with Rhône-Poulenc. The BXN® system allows farmers to spray bromoxynil herbicide directly "over the top" of cotton crops to control bollworms.
3. Dekalb *Bt* Insect-Protected Hybrid Corn is produced by DeKalb Genetics of DeKalb, Illinois, and is engineered to have the *Bt* toxin in the plant tissues themselves. The presence of this toxin confers partial but not complete resistance to major insect pests such as the corn borer.
4. FreshWorld Farms Tomato; Carrot Bites; Mini-Peppers and Cherry Tomatoes are clonally derived plants with longer shelf life and more improved flavor than their grocery market counterparts.
5. Optimum Soybeans, developed by DuPont Agricultural Products, contain oils with higher levels of desirable oleic acid than are found in standard soybeans.

6. High pH Tolerant Corn Hybrids are designed to withstand the typically alkaline soils of some regions in the southern United States. Gray Leaf Spot Resistant Corn Hybrids resist a blight caused by *Cercocopora* species common to the central and southeast corn belt. Both are available through Garst Seed Company.
7. IMI-Corn™ (Imidazolinone) is an herbicide tolerant corn variety developed by American Cyanamid to be resistant to the toxicity of imidazolinone based herbicides including Contour®, Resolve® and Lightning®. In this instance, the genes for herbicide resistance have been developed using natural selection methods. Through 1997, the IMI method has only been transferred to corn seed and no wide-scale planting has been made.
8. Maximizer Hybrid Corn is genetically modified by Novartis Seeds (the parent company based in Basel, Switzerland) to resist the European corn borer.
9. Freedom II Squash from Seminis Vegetable Seeds of Saticoy, California, is genetically engineered to be resistant to plant viruses.
10. The STS® system was designed by DuPont Agricultural Products to be used in conjunction with their Synchrony® and Reliance® herbicides. This system is based on a discovery in 1986 by a DuPont scientist who isolated a genetic trait that enhances the soybean's natural tolerance to sulfonylurea herbicides. DuPont has entered into licensing agreements with four seed companies to incorporate the gene technology into hybrid seed. The STS® system does not use a biotechnology breeding method. The resistant gene is bred from conventional methods.

Large Scale Enterprise

In distinction to these relatively small product ventures, Monsanto Company has embarked on a full scale program in agricultural biotechnology. As shown in Figure 2, Monsanto is currently licensing the gene technology for its bioengineered crops for herbicide or insect resistance to 13 different countries. It presently markets the technology for genes that can convert crops into insect or herbicide-resistant forms. These technologies, known as Roundup Ready™,

Yieldgard®, Bollgard® and BXN®, confer novel properties on the crops they are used in. In its newly released annual report, Monsanto predicts that by 1998 half of the American grain industry will be using their genetically engineered seed. To attain this goal, Monsanto has acquired or controlled at least six of the major seed companies including Holden, Gargiulo Tomato, Naturemark, Asgrow, DeKalb, Delta & Pine Land, Stoneville Pedigreed, and Hartz. The key to Monsanto's success lies in committing the farmer exclusively to their technology.

If this pattern persists, many farming operations will become almost totally dependent on Monsanto Co. for both the chemical and the seed needed each year. From our personal interviews at three such soybean operations in Missouri and Arkansas, farmers may well be tempted to increase the level of the herbicide used to achieve the necessary suppression of undesirable weeds. Coupled with a no-till program, such a cycle might continue until we have the situation analogous to the antibiotic debacle now facing modern medicine: overuse leads to resistant pests. Many evolutionary biologists and entomologists are concerned that what began as a seeming miracle may over time be destroyed by the selection of resistant types, a process that will completely negate any initial advantage to the farmer. To understand how this state of affairs has emerged, we must examine the experiences of the first generation of transgenic seed users.

---

[1] Charles Darwin, *Plants and Animals under Domestication,* London: John Murray, 1868, p. 318.

[2] Quoted in Le Couteur, *On the Varieties of Wheat,* London: John Murray, 1866, p. 16.

[3] Carol Browner, speech given at the Tufts University Medical and Science Center, Boston, Massachusetts, 27 February, 1997.

3

Dangers in Herbicides

In spite of these theoretical concerns, our research and interviews with company officials, United States Department of Agriculture (USDA) and United States Environmental Protection Agency (USEPA) officials, university extension agents, and farmers themselves, show bioengineered food crops gaining widespread acceptance. The most popular bioengineered food crops that have been commercialized contain genes which confer tolerance to a specific herbicide.

To understand the appeal of this innovation, we traveled to the mid-south region of the United States to visit with growers, seed suppliers, and university extension agents in June of 1997. On our visit to Arkansas, enthusiasm was especially evident for herbicide tolerance technologies which permit farmers the greatest possible latitude in using a pesticide. We found glyphosate to be a key pesticide enjoying widespread acceptance. We were surprised to see broadcast aerial spraying of this herbicide on cotton and soybeans, a practice previously limited to pesticides for insect control. One concern focused on the danger of herbicide drift, a view shared by corn growers downwind of transgenic fields who reportedly lost a portion of their non-transgenic crop to drifting herbicides. In part because of this perennial risk, one informant told us he too was going to shift to transgenic crops. His choice was largely predicated on the reputedly excellent safety record of glyphosate, the Monsanto herbicide. But is such blanket optimism justified?

While having a "good" toxicity profile in the 1980s, glyphosate was responsible for more worker-related injuries (at least in California where such records are kept) than all but two other pesticides.[1] We will return to the glyphosate story in Chapter 4. For at least one other herbicide, safety concerns extend to consumers—the people ingesting treated crops—as well as to workers.

The Bromoxynil Story

The rapid commercialization of transgenic cotton provides a key model of how an herbicide with questionable safety can be catapulted into broad acceptance. In this instance, use of a toxic herbicide was greatly expanded by linking it to a new transgenic crop. The herbicide in question is a bromine-containing plant killer made by Rhône-Poulenc known as Buctril®, the brand name for bromoxynil.

In 1995 cotton genetically engineered to be resistant to an herbicide known as bromoxynil was grown from seedstock developed by Calgene, Inc. The end product was marketed as BXN®Cotton. The gene used to make BXN®Cotton was derived from a bacterium that used it to detoxify otherwise toxic nitrile groups (CN) to a relatively benign function known as a carboxylic acid (COOH). In "normal" non-engineered plants, bromoxynil inhibits photosynthesis in plants. By introducing the nitrilase gene into specific crops, transgenic crops are able to detoxify the bromoxynil. As more and more bromoxynil is broken down, a metabolite known as DBHA builds up in the plant tissues.[2] DBHA has been found by Rhône-Poulenc's toxicity testing to carry comparable toxicity to its parent compound.[3]

An Environmental Protection Agency (EPA) registration team assigned to review the possible toxicity of bromoxynil went beyond its mandate and endorsed the newfound ability of the cotton grower to apply bromoxynil *ad libitum* "based on the size of the weed and not to [sic] the size of the cotton" as is usually required.[4] In other words, when the weed problem emerges, bromoxynil can be applied irrespective of growth stage of the transgenic cotton. Before, the cotton plants would be destroyed by virtually any over the top herbicide application.

But the EPA's review gave BXN®Cotton a green light even after finding bromoxynil's by-product in transgenic cotton to be at least as toxic as its progenitor. For unknown reasons the EPA team did not address at least two key considerations. Somewhere in the regulatory shuffle for registering its use on cotton, two glaring realities were overlooked: 1) that herbicide treated cotton and its associated contaminants, while primarily a non-food crop, would nonetheless enter the human food chain. This could happen indirectly through cotton by-products added to livestock feed; or directly, through the production of cotton seed oil; and 2) the DBHA breakdown product of bromoxynil could carry substantial residual toxicity for mammals ingesting BXN®Cotton.

Manufacturers have assumed that the potential human toxicity of the DBHA bromoxynil by-product on cotton would be immaterial since the main product—cotton fabric—is not intended for human consumption. But cotton slash, gin mill leavings and related cotton detritus are widely used in animal foodstuffs, making up to 50 percent of traditional silage. Cotton seed oil is also widely used as a direct human food and cooking additive. In all three forms, we believe residual toxicity from DBHA poses a substantial and largely unmeasured risk. More subtly, cotton dust generated from the processing of BXN® cotton bolls—a major cause of the occupational lung disease known as byssinosis—will be contaminated with residues of bromoxynil and DBHA. The resulting toxicity of cotton dust by this novel form of contamination and any accompanying illness may be exacerbated by toxins in the dust. Neither the Occupational Safety and Health Administration (OSHA) nor the EPA appear to have weighed this possibility in making their safety determinations.

As a result, we are left with troubling unanswered questions. What toxic threat might reside in chemically contaminated dust from transgenic cotton? How much DBHA is found in BXN® derived cotton oils? How much extra toxicity is allowed by the newly established tolerance level in food for BXN® (discussed below) combined with the level permitted in animal feed?

Unfortunately, while bromoxynil's toxicity is fairly well understood, the toxicity of DBHA is largely unknown. The closest approximation is the toxicity produced by its parent compound, bromoxynil itself. According to an EPA Metabolic Activity Committee, both compounds are sufficiently similar in structure to warrant being considered equal in toxicity. But, in fact, the metabolic fate of DBHA in the mammalian body has never been studied—or at least reported. Therefore, DBHA's presence in cotton by-products is a dangerously blank slate since no risk assessments of by-product consumption have been made.

Paradoxically, the risk calculations for human exposure to bromoxynil *exclude* cotton as a significant contributing factor in spite of overwhelming evidence implicating cotton by-products as part of the human food chain. In setting permissible tolerances for bromoxynil, the EPA also made the dubious assumption that only 3 percent of the United Sates cotton crop would contain transgenic cotton, a figure *currently* correct but certainly not the figure set by Rhône-Poulenc and Calgene for the future. If included in the risk equations, bromoxynil resistant cotton would expose us to still more bromoxynil and in so doing, push us over the toxicity threshold. The overall cancer risk from bromoxynil in all *other* food crops is *already* over the threshold for regulatory concern. EPA cancer risk estimates range from 1.5 in a million lifetime cancer risk for bromoxynil contaminated foodstuffs to a total cancer risk of 2.7 in a million for all routes of ingestion or exposure to this chemical alone. For now, the additional food ingestion risk posed by BXN®Cotton has been considered trivial, adding only a trace contribution to the food chain. According to the EPA, at most BXN® could add only another 1 in 10 million risk.

However, as we have pointed out, this risk assumes that only 3 percent of the cotton crop will be transgenic cotton. Additionally, to ensure registration without further review, Rhône-Poulenc moved to reduce the level of any risk, by lengthening the interval between treatment and harvest from 60 to 75 days and lowering the level sprayed to 1.5 pounds per acre. These lowered usage figures were proposed to extend the conditional registration.

We think any EPA decision to permit registration of BXN®Cotton is a major error. To understand why, it is critical to appreciate that even this preliminary registration agreement permits transgenic herbicide resistant cotton to go into full-scale production. We believe the toxicity profile of this chemical makes such a decision a dubious move at best.

Toxicity

Bromoxynil octanoate, the active ingredient in bromoxynil, is converted into bromoxynil phenol (what we have been calling DBHA) when it is metabolized in mammals. Although this step is designed to detoxify bromoxynil and make the molecule more easily excreted by the body, the by-product remains at least as toxic as its parent compound.[5] Additional characteristics of both compounds create other toxicity concerns. Unlike many other phenoxy herbicides which are only soluble in water, bromoxynil and DBHA are soluble in fat, allowing them to bioaccumulate and concentrate within the brain, bone marrow, and other fatty tissues of mammals.

Reproductive damage from bromoxynil or its breakdown products are of particular concern. Bromoxynil is a developmental toxicant which induces structural malformations in mammals when the chemical is administered orally or dermally.[6] A sensitive indicator of developmental toxicity is the induction of supernumerary ribs at very low doses (5mg/kg/day) given during pregnancy.[7] Other forms of developmental toxicity were observed at higher dose levels, including increased overall incidences of minor anomalies; defects in the spine, sternebrae and skull; and reduced fetal weight.[8] These findings clearly qualify bromoxynil as a teratogen. In keeping with this conclusion, dietary exposure in rats caused an increase in developmental disorders in fetuses. Bromoxynil also proved toxic to the female rats in reproductive studies. At sublethal doses, female rats showed severe clinical signs of toxicity including panting, salivation and vomiting.[9] Most critically, the compound produced its birth defects at oral doses *below* those needed for maternal toxicity. In studies conducted using rabbits, bromoxynil also produced birth defects including changes in bone formation in the skull and hydrocephaly.[10] In spite of this disturbing data, in its April 8, 1997 decision the EPA declined to provide a special safety factor to protect infants and young women of reproductive age from bromoxynil residues.[11]

Bromoxynil phenol also has clear-cut carcinogenic activity in at least one animal species. As such, it is classified as a "possible" human carcinogen by the EPA. Feeding studies have shown that bromoxynil phenol produces an increased incidence of liver tumors in both female and male mice.[12] Two studies have supported the evidence of this increased carcinogenicity. A 1982 study observed increases in liver tumors at doses of 30 parts per million and 100 parts per million.[13] In 1994, Rhône-Poulenc sponsored a carcinogenicity study which also detected increases in adenomas (benign tumors of glandular origin) at dietary levels of 75 ppm and 300 ppm and carcinomas at 20 ppm and 300 ppm.[14]

Regulatory Information

Initially, the EPA found this data to lack compelling force to limit bromoxynil's registration. On May 5, 1995, the EPA granted Rhône-Poulenc conditional registration for the use of bromoxynil with BXN® Cotton. Disregarding the health and safety concerns, the EPA renewed its use permit and established a temporary tolerance. This conditional registration was set to expire on April 1, 1997.

On July 25, 1996, Rhône-Poulenc requested an extension of the time limited tolerance for bromoxynil on transgenic cotton. Rhône-Poulenc requested that the tolerance be kept at .04 ppm even though it failed to present, in our opinion, conclusive data on the safety profile of the DBHA metabolite.[14] We believe Rhône-Poulenc may have acted preemptively to ensure the acceptability of its transgenic seed for the 1997 planting season. In an apparent concession to the EPA, Rhône-Poulenc most likely sought a low tolerance (at .04 ppm) which would still permit farmers to buy the transgenic variety. This entry level decision was critical to any long-term strategy. Were cotton growers to purchase the conventional varieties that are less expensive, Rhône-Poulenc would lose an entire growing season and perhaps the "brand loyalty" they needed to ensure future reliance by cotton growers on their new transgenic variety. (Most farmers begin buying their seed for the next seasons during the winter prior to planting.)

In a major concession to Rhône-Poulenc, the EPA issued a Notice of Proposed Rulemaking on May 2, 1997, which established permissible tolerances for residues of bromoxynil and the metabolite DBHA on cotton commodities. The proposed tolerances were limited to one year, expiring January 1, 1998, and farmers were limited to planting a maximum of 500,000 acres or 3 percent of the nation's cotton acreage. The EPA's proposed tolerances for undelinted cottonseed were set at 7 ppm; for cotton gin by-products at 50 ppm; and for cotton hulls at 21 ppm (US EPA, 1997).[15] Most interestingly, these tolerances were reduced to fit the manufacturer's "new" application rates and extended pre-harvest intervals. Such a move by

Rhône-Poulenc was perhaps, in our opinion, a preemptive strike to ensure any risk assessment would be based on the new tolerances. Our rough calculations show that bromoxynil would survive a risk assessment if kept to these levels, coming in at just *under* the threshold for regulatory action.

In spite of this regulatory uncertainty, the EPA permitted 3% of the United States cotton to be BXN®Cotton in 1997. Even as they gave this okay, the EPA determined a permanent tolerance could not be established for bromoxynil on cotton due to major data gaps in toxicity testing.[16] These gaps were uncovered by the EPA while reviewing studies submitted by Rhône-Poulenc. The EPA believed that the most significant data gap involved the absence of a testing method and residue data for the metabolite DBHA in cottonseed and cotton gin trash.[17] We find this omission troubling since this metabolite accounts for most of the toxic residue found in BXN®Cotton.

Comment

The EPA's decision to permit continued registration appears to be motivated as much by practical considerations as by toxicity concerns. Present tolerances do take into account existing residues of DBHA *and* the parent compound in cottonseed and cotton gin by-products.[18] They also allegedly take dietary risk into account, although no cotton oil is assumed to be contaminated. With these parameters, the carcinogenic risk from bromoxynil is supposedly "negligible" within the meaning of standard risk assessment.[19] In a kind of regulatory double-standard, the EPA asserts the actual lifetime upper-bound carcinogenic risk for bromoxynil (1.5 in a million) and from all sources (2.7 in a million) are "well within" the 1 in a million risk standard for regulatory purposes.

We question if the 1.5 in a million level technically meets the requirements of the new Food Quality Protection Act (FQPA). This act amended Section 408 of the Federal Food, Drug and Cosmetic Act (FFDCA) which authorizes the establishment of tolerances, exemptions from the requirement of a tolerance, modifications in tolerances and revocations of tolerances for residues of pesticide chemicals in or on raw agricultural commodities and processed foods. The FQPA created a new subsection to Section 408 that clearly states the EPA must determine if any given tolerance is "safe." "Safe" in this instance is defined as having "reasonable certainty that no harm will result from aggregate exposure to the pesticide chemical residue, including all dietary exposures and all other exposures for which there is reliable information."[20] The benchmark for safety is one in a million. In the EPA's view, the proposed tolerances for the 1997–98 growing season were safe within the risk assessment uncertainty and the parameters of the Food Quality Protection Act.

We take issue with this conclusion for bromoxynil on four counts: First, the aggregate food source risk from cotton is already 1.5 in a million, which is above the regulatory threshold.[21] Second, the combined food and non-food risks are *2.7* in a million. Third, the risk assessment fails to factor in possible bioaccumulation of the herbicide if cattle are left "on feed" for protracted periods. Fourth, it neglects interactive effects that might increase bromoxynil's reproductive toxicity. Such effects could occur, for instance, if a person were concurrently exposed to other teratogens like phthalates in foodstuffs, other pesticides or chemical toxicants during pregnancy.

In our view, these considerations for BXN® Cotton clearly push overall bromoxynil exposures over the safety threshold. We also believe the EPA's original tolerance for BXN® was motivated by the preexisting regulatory laxity towards transgenic crops we have documented in other agencies. Additionally, no monitoring program expressly designed to measure bromoxynil residues is in place. As a result, we are gravely concerned. Consider how we reached this precarious endpoint: Rhône-Poulenc pressed its BXN® seed into production *before* the DBHA residue "data deficiencies" had been fully assessed.[22] If registration is permitted, consumers of cottonseed oil or bromoxynil contaminated meat may be at high risk of assimilating unmonitored toxic residues. For a few people, perhaps the very old, infirm or young, the exposure may be sufficient to harm. Should this occur, it will be the first instance of transgenic crop poisoning in the U.S. While the EPA had enough misgivings initially to deny Rhône-Poulenc an extension of temporary tolerances for bromoxynil on BXN®Cotton for the 1998 growing season,[23] it subsequently relented, permitting 10% of 1998's cotton crop to use the BXN® variety. A second high-risk venture will also be played out in the ensuing 2–3 years: the widespread reliance on Roundup Ready™ soy products.

[1] Pease, W.S., et al. 1993. "Preventing Pesticide-Related Illness in California Agriculture: Strategies and Priorities." Environmental Health Policy Program Report. Berkeley, CA. University of California, School of Public Health. California Policy Summary.

[2] R. Griffith, K. Boyle, L. Hansen, G. Kramer and S. Robbins, United States Environmental Protection Agency, Office of Prevention, Pesticides, and Toxic Substances. "Extension of Conditional Registration for the Use of Bromoxynil on Transgenic Cotton," memo to D. Stubbs dated 8 April, 1997.

[3] Griffin, R. USEPA. Office of Prevention, Pesticides, and Toxic Substances. "Characterization of Food/Water Carcinogenic Risk Estimates for Bromoxynil," #PP3F04233. 16 April, 1997.

[4] R. Griffith, et al. Memo to D. Stubbs dated 8 April, 1997.

[5] United States Environmental Protection Agency, Office of Prevention, Pesticides, and Toxic Substances, "Bromoxynil Phenol and Bromoxynil Octanoate; Toxicology Branch Chapter for the Reregistration Eligibility Decision Document (RED)," 12 March, 1997.

[6] United States Environmental Protection Agency, Office of Prevention, Pesticides, and Toxic Substances, "RfD/Peer Review of Bromoxynil (phenol): 3,5-dibromo-4-hydroxybenzonitrile," 12 April, 1996.

[7] United States Environmental Protection Agency, Office of Prevention, Pesticides, and Toxic Substances, "RfD/Peer Review of Bromoxynil (phenol): 3,5-dibromo-4-hydroxybenzonitrile," 12 April, 1996.

[8] United States Environmental Protection Agency, Office of Prevention, Pesticides, and Toxic Substances, "Bromoxynil Phenol: Identification of New Toxicological Endpoint of Concern for Acute Dietary Risk Assessments," 3 April, 1997.

[9] Hazardous Substance Data Base, 1992, National Library of Medicine, Toxnet, Bromoxynil.

[10] United States Environmental Protection Agency, Office of Prevention, Pesticides, and Toxic Substances, "RfD/Peer Review of Bromoxynil (phenol): 3,5-dibromo-4-hydroxybenzonitrile," 12 April, 1996.

[11] United States Environmental Protection Agency, Office of Prevention, Pesticides, and Toxic Substances, "Carcinogenicity Peer Review of Bromoxynil Phenol." 12 March, 1997.

[12] United States Environmental Protection Agency, Office of Prevention, Pesticides, and Toxic Substances, "Carcinogenicity Peer Review of Bromoxynil," 14 February, 1996.

[13] United States Environmental Protection Agency, Office of Prevention, Pesticides, and Toxic Substances, "Carcinogenicity Peer Review of Bromoxynil," 14 February, 1996.

[14] Loranger, Richard. USEPA. Office of Prevention, Pesticides, and Toxic Substances. "Bromoxynil on Transgenic Cotton. Request for Conditional Registration and Time Limited Tolerance," CB#14981. 26 January, 1995.

[15] United States Environmental Protection Agency, Office of Prevention, Pesticides, and Toxic Substances, "Bromoxynil Tolerances; OPP-300486," 2 May, 1997.

[16] Loranger, Richard. USEPA. Office of Prevention, Pesticides, and Toxic Substances. "Bromoxynil on Transgenic Cotton. Request for Conditional Registration and Time Limited Tolerance," CB#14981. 26 January, 1995.

[17] Loranger, Richard. USEPA. Office of Prevention, Pesticides, and Toxic Substances. "Bromoxynil on Transgenic Cotton. Request for Conditional Registration and Time Limited Tolerance," CB#14981. 26 January, 1995.

[18] United States Environmental Protection Agency, Office of Prevention, Pesticides, and Toxic Substances, "Bromoxynil Tolerances; OPP-300486," 2 May, 1997.

[19] United States Environmental Protection Agency, Office of Prevention, Pesticides, and Toxic Substances, "Bromoxynil Tolerances; OPP-300486," 2 May, 1997.

[20] Food Quality Protection Act, Section 408 (b)(2)(A), 104th Congress, 1996.

[21] R. Griffith, K. Boyle, L. Hansen, G. Kramer and S. Robbins, United States Environmental Protection Agency, Office of Prevention, Pesticides, and Toxic Substances. "Extension of Conditional Registration for the Use of Bromoxynil on Transgenic Cotton," memo to D. Stubbs dated 8 April, 1997.

[22] Kramer, G. F. Health Effects Division, Office of Prevention, Pesticides, and Toxic Substances, USEPA, "Bromoxynil Octanoate and Heptanoate in or on Transgenic Cotton. Amendment of 12/18/96. Submission of Residue Data," 11 March, 1997.

[23] Laws, Forrest. "EPA Denies Buctril New Tolerance for '98," *Delta Farm Press.* 9 January, 1998:1–2.

4

Are We Ready for Roundup Ready™ Foods?

"The biggest mistake that anyone can make is moving slowly, because the game is going to be over before you start."

—Hendrik Verfaillie, Senior Vice President and Chief Financial Officer, Monsanto Company

The Soybean Story

We have already documented how soybeans are being commercially released with newly inserted genes making them tolerant to herbicides. The history of our increased reliance on this once exotic crop is illuminating. About three thousand years ago, farmers in China began planting the black and brown seeds of a wild recumbent vine. The crop being cultivated was the forbearer of modern-day soybeans. The reasons why Chinese farmers chose to cultivate soybeans remain ineffable. In its original form, the soybean spread out along the ground making it difficult to cultivate and harvest. Its seeds or beans were hard and indigestible.[1] By 1100 B.C., the plant had been trained through genetic selection to stand upright and bear larger, more useful seeds. The beans were valuable, producing plentiful protein while growing in soils too depleted to support other crops. By planting this crop, farmers also discovered soybeans enriched the soil. We now know this benefit comes from nitrogen-fixing bacteria in soybean root nodules that transform N_2 into ammonia.

Over the next thousand years, the soybean became a revered staple food of the Chinese people. Different varieties carried names like Great Treasure, Brings Happiness, Yellow Jewel and Heaven's Bird.[2] The soybean was used in the production of soy nuts, soymilk, soy sauce, miso (fermented soy paste), tempeh (fermented soy cake), flour, doufu (tofu in Japan) and soybean oil. Because of its high percentage of protein, tofu became an especially integral part of the strict vegetarian diet required of Buddhist monks.

Origins

The soybean is a legume from the Phaseolea family. The cultivated variety of soybean is the genus *Glycine max.* The wild and weedy forms of the soybean, *Glycine soja* and *Glycine gracilis,* are found in China, Korea, Japan and Russia. *Glycine max* is sexually compatible with these relatives, although according to the United States Department of Agriculture (USDA), there is little threat of the wild and domesticated plants interbreeding because the wild relations do not grow in North America.[3] *Glycine max* has been hybridized by seed producers in the United States. Characteristics are bred into the plant by natural means to increase its vigor and yield. The soybeans are planted each year to produce viable seed due to the plant's ability to self-pollinate.

The soybean was introduced into the West in 1765 by Samuel Bowen, a merchant who brought a few hundred seeds back from his travels to China. The bean was prized for its nutritional quality over the next several decades, although it did not become a part of the human diet for the New World people until after the middle of the 19th century.[4] The plant was originally grown as fodder or animal feed. In the 1930s, Dr. Fearn brought a low temperature cooking method for de-bittering soybeans from China to the United States. Gradually, the production of soybeans increased throughout the United States because it was so easy to grow and it lacked the acreage restrictions such as those in place for corn and cotton in the 1930s during the New Deal period. Soybean production continued to grow as its versatility as a food supplement (in the form of lecithin) and its utility in cooking (in the form of soybean oil) were recognized.

Why Engineer Soybeans?

Given its remarkable nitrogen fixing activity and its natural competitive advantage over most weeds, why was it necessary to change its genetic capabilities so radically to tolerate herbicide application? The answer to this and related questions of transgenic production turns on enormous profit margins possible through even modest increases in soybean production.

The United States, Brazil, China and Argentina produce 90% of the world's soybeans. The United States alone produces 47% of this supply and exports one-third of its production. Soybeans provide a major source of vegetable oil and high-protein feed supplements for livestock. The oil pressed from soybeans accounts for 80% of the vegetable oil consumed in the United States.[5] Such oil is used in many common products such as margarine, mayonnaise, shortening and salad dressings.

Soy protein, valuable as it is for human consumption, is largely passed over by food processors and fed to animals. This extraordinary waste can only be understood in light of the peculiarities of American and European consumer preferences. Ninety to ninety-five percent of the soy meal, including the outer membranes (hulls) of the beans, continues to be used for animal feed, where it enhances meat production.

Scale of Production

In 1997, of the 60 million acres of the *Glycine max* soybeans grown in the United States, 10 million acres, or 15% of this total were genetically altered to resist the herbicide Roundup®. The seeds needed to achieve this property are marketed as Roundup Ready™ and are sold through seed distributors such as Hartz and Asgrow Seed Company.

Ironically, we can infer the Roundup Ready™ genetic technology used to transform seed has been funded by the profits from the Roundup® herbicide. The patent Monsanto acquired for Roundup® expires in the year 2000. With the demise of sole ownership of this compound just around the corner, Monsanto has devised a clever strategy for keeping growers dependent on their herbicide. By contract with Roundup Ready™ seed-buying growers, Roundup® is the only herbicide that can be used with seeds created with the Roundup Ready™ technology. Farmers are obliged to make a commitment to exclusive Roundup® use as a condition for buying Roundup Ready™ seed [see box].[6]

> *The Grower agrees not to supply any of this seed to anyone for planting and agrees not to save any crop produced from this seed for replanting or supply saved seed to anyone for replanting. The grower agrees not to use this seed or provide it to anyone for crop breeding, research or seed production. If a* [sic] *herbicide containing the same active ingredient as Roundup® Ultra herbicide (or one with a similar mode of action) is used over the top of Roundup Ready™ soybeans, the Grower agrees to use only the Roundup® branded herbicide.*
>
> Source: Purchase Log Report Form, 1997 Hartz™ Seed Company, A Unit of Monsanto.

A Closer Look at Roundup® and Roundup Ready™ Products

In 1992 Monsanto scientists completed their preparation of Roundup Ready™ technology. Roundup Ready™ plants contain a special gene isolated from either a petunia or a bacterium.[7] This gene protects plants from Roundup® by overproducing the natural levels of the key enzyme normally poisoned by the herbicide. This critical enzyme assists the photosynthesis pathway that lets plants make sugars, amino acids, and ultimately proteins from the energy in sunlight.[8] By forcing the plant to overproduce the essential enzyme, Roundup Ready™ technology permits transgenic plants to increase levels of the enzyme in their chloroplasts, dampening, but not eliminating Roundup's® toxic effects. The Roundup Ready™ technology provides a kind of genetic buffer against the toxic effects of Roundup® to the chloroplast. (Chloroplasts are the chlorophyll-containing photosynthesizing organelles of plants thought to be the descendants of endosymbiotic cyanobacteria.)

Toxic Concerns

A logical concern over Roundup Ready™ technology is that its widespread use will go hand and glove with a dramatic increase in

Roundup® usage. In spite of its reputation as a "safe" herbicide, we have concerns over this vast expansion of reliance on Roundup®. While Roundup's® major ingredient, glyphosate, has toxicity of its own (see below), toxicity from its other components add another dimension to the debate. Roundup® itself is typically only 41% glyphosate. The remaining 59% consists of inert ingredients. These "inert" ingredients are not inactive. They are "inert" only with regard to their toxicity to pests. Inerts like polyethyloxylated tallow amine surfactant (POEA) have toxicity of their own in their function as detergents or surfactants whose purpose is to ease the application of the herbicide. Surfactants de-clog applicators and evenly spread the herbicide over the plants.

One notable inert ingredient in Roundup® is POEA. POEA was cited as the cause of toxicity in nine Japanese human deaths after ingestion of Roundup®.[9] In Taiwan, deaths have been reported in some 11 of 97 persons who have intentionally ingested large amounts of glyphosate and POEA.[10] While such doses (measured in hundreds of milliliters) are not likely to be encountered in everyday usage, they confirm the toxicity of glyphosate surfactant herbicide formulations. Thus, a major question is whether Roundup Ready™ technology increases toxic risks by greatly enhancing the tonnage of chemical usage worldwide.

Toxicity of Glyphosate

After being fed to test animals at high levels (hundreds of parts per million), glyphosate has been known to cause increases in bile acids, alkaline phosphatase and alanine aminotransferase, all of which suggest toxicity to the liver and its detoxifying system.[11] Roundup® may also damage non-target plants. After treatment with Roundup®, some increases in chromosome aberrations have been reported but may be due to the toxicity of the mixture or its surfactant component.[12] Roundup® may also produce secondary effects not considered in the initial risk assessment. For instance, high doses of glyphosate can inhibit the monooxygenases mammals needed to detoxify other chemicals, an effect the authors attributed to possible disruption of the cellular membrane.[13]

In support of glyphosate's likely environmental safety, industry toxicologists emphasize that the chemical is water soluble and is likely to break down in the environment without significant bioaccumulation. True, glyphosate binds tightly to soil particles and is subject to rapid environmental degradation. We concur that, being water soluble, Roundup® is unlikely to bioaccumulate in the food chain as do DDT, heptachlor, and other fat-soluble molecules. But the chemical does build up in the hulls of the soybeans used in animal feed. While it is unlikely that humans would ingest toxicologically significant levels of glyphosate were they to eat soybean oil from transgenic crops or milk from animals fed soy byproducts, they still might accumulate significant amounts of glyphosate residues. In particular, heavy meat eaters might have exposure from meat and animal byproducts that become contaminated in the course of feedlot programs using Roundup Ready™ soybeans. This is because livestock can be fed soybean hulls, and the bacteria that thrive in the intestinal tract metabolize glyphosate into a more fat soluble and toxic amine that could, in theory, accumulate in their body tissues.

Genetic Transformation

Through genetic engineering, soybeans have undergone an unprecedented genetic metamorphosis. They have acquired novel chemical properties without any change in their gross physical features. These changes mark a revolutionary point of departure for agriculture. Traditional, timeworn selection and growing methods have been upstaged by fast track programs developed by the biotechnology industry. Manufacturers bent on rapid "progress" have achieved these changes in a single generation instead of the decades previously required.

By genetically tying crops to certain chemicals, the industry is also committing a generation of farmers to a new form of dependency. Transgenic technologies are moving farmers away from integrated pest management and plant protein improvement programs into straight herbicide-driven economies of scale. In 1997, two-thirds of the bioengineered crops on the market or proposed for use were created for herbicide tolerance. If successful, this new revolution will commit a generation of farmers to continued use of company-selected herbicides.

Two Case Studies in Herbicide Tolerance

We visited several such farmers in the Southeast United States at the beginning of the 1997 growing season to gain better appreciation of the impact of this technology. One farmer was Mr. George Zanone, of Horseshoe Lake, Arkansas. Mr. Zanone has lived in Arkansas all of his life. He is an affable gentleman in his 50s, given to khakis and plaid shirts. Farming has been in his family for more than three generations. Mr. Zanone watches the commodities market daily. He is in touch with the latest agricultural technologies that allow him to stay competitive. Mr. Zanone has been working the family farm for over 20 years. He watched his grandfather farm the same 8000 acres. Presently his son, George, Jr., is learning the skills needed to take over his business. In the past two years, Mr. Zanone told us he had acquired as much Roundup Ready™ seed as he could obtain from his seed distributor. In 1997, he planted over half of his acreage using bioengineered seeds. We asked him why. Over a three hour interview he showed us his luxuriant and verdant fields of cotton and soybeans, and comfortably explained the economic advantages acquired by using the Roundup Ready™ program. For Mr. Zanone, Roundup Ready™ was clearly an absolute wonder, an economic miracle. At one point we thought we were in the company of a Monsanto detail man.

As we watched him survey his computer screen for commodity futures, he itemized the advantages of Roundup Ready™ technology to his operation. Roundup Ready™ technology allows Mr. Zanone to farm his 8000 acres with less than a dozen people. Before the expansive use of agricultural chemicals, 20 years ago, he remembers needing ten families to farm just 800 acres. Prior to using Roundup®, he applied herbicides four to five times during a season, whereas the latest technology requires that Roundup® be applied only twice a growing season. This decrease in herbicide and labor needed for his crops means more money for Mr. Zanone in a given growing season. According to Mr. Zanone, Roundup Ready™ seed gives him the same yield at a lower cost. Every morning he watches the stock market's opening and closing costs on agricultural commodities and thanks Monsanto for their technology because he is saving close to $20.00 per acre on herbicide. With this new competitive edge, he feels like he can finally compete with larger farmers.

We looked down the vast rows of his cotton and soybean crops in June of 1997. We saw Roundup Ready™ fields. In many places the soy and cotton were grown side by side, as they both could be sprayed with the same herbicide—and would tolerate its otherwise blistering effects. He showed us brown and withered weeds between the rows and marveled at the cotton and soybean plants' tolerance. The fields were considered "clean," and the plants were beautiful. Not a worker was in sight.

Another farmer, Mr. Lowell Taylor, is about 70 years old and has been farming soybeans and cotton next to Horseshoe Lake, Arkansas, all of his life. A genteel man, he was reluctant to tell me just how large his farm was. We sat down at his dining room table and were served an enormous lunch of fried green tomatoes, pork chops and homemade apple crisp for dessert.

Mr. Taylor does not use Roundup Ready™ products at all. He does not believe in them. Like most old-time farmers, he is ingenious. As Mr. Taylor says, put a farmer in charge of his crops and he will figure out any way possible to make a living. In part, Mr. Taylor has done this through saving his own seed from year to year. It takes approximately 50 pounds of seed for every acre planted. Most seed costs around $15.00 per 50 pound bag. In 1997, Roundup Ready™ seed retailed for more than twice that much and in 1998 probably three times that much, not including the Monsanto "technology fee."

The "technology fee," an additional cost of $8.00 per acre for the Roundup Ready™ products, really rankled Mr. Taylor. In theory, all of it goes directly back to Monsanto to develop more genetic technology or pay for its original production costs. Mr. Taylor may not know that Monsanto has probably been repaid *many* times over for its technology through increased sales of Roundup® herbicide.

When a farmer buys Roundup Ready™ seed, he must sign an agreement stating he will not save any engineered transgenic seed at all for the next year's planting. For a seed saver like Mr. Taylor, not being able to save seed confirmed his decision that he would not use the product. In the future, if Monsanto is successful and acquires full control of the soybean seed market, farmers like Mr. Taylor may have even less choice over using Roundup Ready™ seed.

The Seed Company

Hartz™ Seed Company is a unit of Monsanto Company located in Stuttgart, Arkansas. They are the second largest soybean seed distributor in the United States. Since the United States is the largest soybean grower in the world, it may be safe to say that Hartz™ is the second largest soybean seed company in the world. Their facilities are located in what is referred to as the "Grand Prairie": 50 square miles of valley where record yields of rice and soybeans have been grown.

The seed plant looks like a huge ship sitting on an ocean of green. Ten enormous silos house close to two million pounds of seed. Beginning in 1996, Hartz™ began distributing Roundup Ready™ seed to thousands of farmers in the mid-south region, which includes Mississippi, Tennessee, Arkansas, Missouri and Louisiana. Of the 60 million acres of soybeans grown in the United States, close to ten million of those acres in 1997 will contain the gene expressing tolerance for Roundup® herbicide.

When we spoke with them in the early summer of 1997, Hartz™ staff members Keith Thompson, Vice President and General Manager, and Larry Spooner, Vice President of Sales, emphasized that their company is predicting that by the year 2000 the entire Hartz™ operation will be based exclusively on Roundup Ready™ products. Cryptically, they pointed out that if farmers do not want to use these products after the millennium, they will have to go elsewhere for their seed. Thompson re-emphasized how much crop will be devoted to transgenic technology: 50% by 1998, to 75% in '99, to 100% by the year 2000.

Hartz™ Seed is working especially aggressively in South America where they are distributing seed to Argentina. Not coincidentally, Monsanto spent $80 million on a glyphosate facility in Zarate, Argentina, in 1997. Monsanto has recently bought a large soybean distributor in Brazil by the name of Monsoy. This company will be operated under the auspices of Hartz™, sharing germ plasm and selling the product throughout Brazil. Monsoy and Hartz™ have worked together since the buyout, engineering the technology to produce the proper seed for the Brazilian climate. This has taken place in spite of the fact that the people of Brazil are reluctant to adopt the technology, and as of late 1997, the government had not yet passed regulations for the product. Mr. Thompson described their reluctance as "bull-headed."

"The Brazilians just want to reinvent the wheel" regarding regulations, he declared. In distinction, he said, "the Argentines just bought the whole package." Monsanto, Hartz™ and now Monsoy are simply waiting for the regulations to allow them to begin the Brazilian distribution. There is no doubt in the minds of the gentlemen from Hartz™ that their product will be allowed. It is just a matter of time.

Hartz™ also provides seed to the Japanese market, although this is not for planting purposes. They provide a Japanese company with non-transgenic seed as a raw commodity to be used in making soy products such as miso and tofu. Presently, the people of Japan are very suspicious of any products that are bioengineered. Hartz™ has informed them that there are no safety issues with the Roundup Ready™ soybeans. They have also given notice to the Japanese that after the next few years, engineered soybeans may be all they will be offered. Hartz has taken this position in spite of the fact that the Japanese eat the entire bean in their traditional diet. Current studies that have been submitted to the EPA for food safety evaluation include only tests for residue in soy oil and not the entire bean.

The Extension Agent

Dr. Ford Baldwin is an extension agent and weed scientist working out of the University of Arkansas. He works with approximately 250 farmers throughout the state. Dr. Baldwin says that he has received more complaints regarding pesticide drift this year (1997) than any of the 20 years in his field. In large part he believes this is due to the mass use of Roundup Ready™ seed and concurrent increase in spray acreage. Pesticide-drift caused crop destruction increases the pressure for non-users like Mr. Taylor to get on the Roundup Ready™ bandwagon. A neighboring farmer not using the technology stands to have his crops destroyed if the Roundup® herbicide drifts onto his fields.

A key factor increasing herbicide drift is the type of surfactant currently selected by Monsanto for use with the herbicide. A surfactant is added to herbicides to allow ease in application from the spray nozzles. Traditionally, surfactants are added by the company that retails the herbicides. This allows the small retailers the opportunity to make a few extra dollars per gallon of herbicide. The retailers can no longer do this with Roundup®. Monsanto is adding a pre-mix of their own (POEA) to the Roundup® Ultra herbicide, the same toxic ingredient we highlighted at the beginning of this chapter.

Many farmers and scientists believe a major reason for the increasing drift problem is that the Monsanto surfactant sprays more finely, which increases the likelihood of drift. Dr. Baldwin's view is that "in the real world we are drifting Roundup® all over the damn state of Arkansas."[14]

What Does It All Mean?

Presently farmers have a choice to use Roundup Ready™ seeds or traditional seeds with the same genetic background—and not the improved varieties being developed at several university extension centers we surveyed. In the next two years, the risks involved if a farmer does not choose Roundup Ready™ technology will increase exponentially. Not only will the farmer run the risk of having his crop drifted over by Roundup® (which could kill his crop), but he will be at a competitive disadvantage with farmers who may be saving $20.00 per acre on chemical costs and purportedly achieving the same yields.

Clearly, Monsanto has established a corner on the agricultural market for soybeans by controlling the chemical, seed and the distribution channels to the farmer. Everyone seems to be making money and everyone is happy. So what is the problem?

We have two major concerns: this technology is creating a reliance upon herbicides whose long-term effects remain largely unexamined. We now know some chemicals can damage ecosystems invisibly by affecting the soil's microflora. Roundup® is no exception. It kills some soil microorganisms outright, notably a type of fungus *Penicillium funiculosum*, and stimulates others, disturbing microbiological equilibria.[15] Second, we see an ominous potential for Monsanto to drive its competitors out of business, thereby limiting the choices available to soybean farmers and consumers. This may create a monopolistic situation where an entire commodity will be controlled by one company. In 1997, in our view, Monsanto was well on its way to complete domination of at least soybean agriculture.

We have made several attempts to discuss the spectrum of our concerns about Roundup Ready™ technology with Mr. Robert Shapiro, now ex-CEO of Monsanto Company. He has consistently declined to be interviewed by us in this regard.

Our View

In spite of claims from Monsanto executives that Roundup Ready™ technology will feed the world, we believe the advantages of using Roundup Ready™ seed are purely economic. Using this technology means more money for the farmer and more money for Monsanto. As with the toxic chemical bromoxynil, it is unlikely we will learn of any of the adverse effects to human health before it is too late to do much about them. All of the field studies submitted to the USDA for these plants were done on less than ten acre plots. We have no idea how these herbicides will act when sprayed over millions of acres of transgenic cotton, corn, or soybeans planted on United States soil. At least one adverse event, the loss of approximately 30,000 acres of transgenic cotton (described in the subsequent chapter) may be the tip of an iceberg of hidden consequences. An equally disturbing prospect awaits the planned expansion of pest resistant crops. In this instance, the concern is how mass reliance on a single breeding technology threatens ecological stability.

---

[1] Hapgood, F. "Soybean," *National Geographic.* 172: 76–91, 1987.

[2] Hapgood, 75.

[3] "Biology of a Soybean," USDA/Animal Plant Health Inspection Service report, 1996, 3.

[4] Hapgood, p. 79.

[5] Frank Flighter, edible oils consultant, United Soybean Board, Personal interview, 26 February, 1997.

[6] L. Spooner, Personal interview, 20 June, 1997.

[7] Federal Register. Volume 58, No. 232. Monday December 6, 1993. Receipt of Petition for determination of nonregulated status of genetically engineered soybean line. The donor gene comes from a bacterium called *Agrobacterium* sp., strain CP4, and codes for a protein enzyme consisting of a polypeptide of 455 amino acids known as 5-enolpyruvylshikimate-3-phosphate synthase (EPSPS).

[8] Federal Register Notices, 93-148-1, Vol. 58: 232, 6 December, 1993.

[9] Sawanda, Y., Y. Nagai, M. Ueyama, and I. Tamamoto et al. "Probable toxicity of surface active agent in commercial herbicide containing glyphosate." *Lancet* 1: 299 (1988).

[10] R. L. Tominack, G. Y. Yang, H. M. Chung, J. F. Deng. "Taiwan National Poison Center survey of glyphosate-surfactant herbicide ingestions," *Journal of Clinical Toxicology* 29(1): 91–109, 1991.

[11] U.S. Department of Health and Human Services, Public Health Service, National Institutes of Health, NTP Technical Report on Toxicity Studies of Glyphosate (CAS No. 1071-83-6) Administered in Dosed Feed to F344/N Rats and B6C3F Mice. (NIH publication 92-3135). Toxicity Report Series No. 16. Research Triangle Park, NC: National Toxicology Program.

[12] J. Rank, A. G. Jenson, B. Skov, L. H. Pederson and K. Jenson. "Genotoxicity testing of the herbicide Roundup™ and its active ingredient glyphosate isopropylamine using the mouse bone marrow micronucleus test, Salmonella mutagenicity test and Allium anaphase-telophase test," *Mutat. Res.* 300(1): 29–36, 1993.

[13] E. Hietanen, K. Linnainmaa, H. Vainio. "Effects of phenoxyherbicides and glyphosate on the hepatic and intestinal biotransformation activities in the rat," *Acta. Pharmacol. Toxicol.* 53: 103–12, 1983.

[14] Baldwin, Ford, University of Arkansas Cooperative Extension Agent, Personal interview, 19 June, 1997.

[15] Abdel-Mallek, A. Y., Abdel-Kader, M. I. A., Shonkeir, A. M. A. "Effect of glyphosate on fungal population, respiration and the decay of some organic matters in Egyptian soil," *Microbial Research* 149: 69–73, 1994.

5

Destroying a Miracle

Crops engineered with genes for resistance to herbicides are but one commodity in the transgenic enterprise. In addition to these herbicide tolerant crops, others are being created to be intrinsically insect resistant. This seemingly ingenious ploy has been achieved by isolating and transplanting genes that make a plant noxious to its pest. In effect, genetic researchers have figured out how to put a "poison pill" into every cell of a crop plant, hopefully without making it poisonous to people. To do so, transgenic scientists have usurped a "natural" solution to crop infestations. Corn, cotton and potatoes are being engineered to contain genes from a particular bacteria known as *Bacillus thuringiensis* (*Bt*) and related species which also produce insect-damaging toxoids.

Historically *Bt* bacteria have been used in integrated pest management (IPM) strategies because of their remarkably selective toxicity for the leaf-eating larvae of moth and butterfly species. *Bt* bacteria organisms contain a crystalloid toxin that carries insecticidal properties that is activated only in the uniquely alkaline environment of the intestinal tract of coleoptera or lepidoptera larvae. By isolating the gene for this toxin, scientists believe they have captured a natural biocide for commercial use. If nothing else, *Bt* toxoid—if properly promoted—promises major returns for its management team. Commentators estimate the worldwide market for *Bt* based commodities to be $25 billion.[1]

Monsanto Company has been at the forefront of the *Bt* movement. They have been granted patents on their *Bt* gene technology for use in Bollgard® (cotton), Yieldgard® (corn) and New Leaf® (potato) products. Currently Bollgard® cotton accounts for approximately 6 million acres of the 14 million acres of cotton grown in the United States. Yieldgard® corn is expanding into the U.S commercial market for the first year. In 1997, it will comprise 6 million of the 80 million total acres of corn. New Leaf® potatoes have not yet been introduced in mass quantity. In 1998, Monsanto plans to expand their *Bt* market worldwide by introducing their *Bt* containing products to China, South Africa, Mexico and Zimbabwe.[2] Other companies are also marketing *Bt* engineered products in Europe, including Ciba-Geigy and Novartis.

At first glance, this global movement appears promising from an ecological viewpoint, in large part because it would theoretically reduce reliance on chemical pesticide control. At another level, the mammoth movement of genes for *Bt* into dozens of different crop plants is the forerunner of a massive evolutionary re-shifting. Dozens of strains of insect pests will find literally every bite of their plant diet contaminated by a toxoid protein product. Some of those that do not die will almost certainly contain toxoid resistance genes. Others will be selected to vary their crop preference, sending hordes of insects into neighboring ecosystems for forage. Over time, perhaps just one or two growing seasons, what has been an ecological miracle may become an ecological disaster.

History

We might reasonably ask how this formidable invention came into existence. *Bacillus thuringiensis* was discovered in Thuringia, Germany, in 1911, from one of a multitude of common soil bacteria. It has been available in commercial formulations for insect control since the 1930s, yet it remained a minor component of pest management until recently.[3] Only a few of the *Bacillus* strains contain insecticidal characteristics. Until the 1980s, strain *kurstaki* was the only breed in widespread use as a pest control agent. It was widely used for controlling lepidopteran larvae. Currently, new strains are being discovered and rapidly developed. An example of a newly developed strain is *Bt tenebrionis* that is effective against the Colorado potato beetle.

How Does It Work?

Bacillus thuringiensis produces crystallized toxins that disrupt the digestive system of the target insect. The insect eats the leaf containing the bacteria. The bacteria proliferate in the alkaline environment of the insect's gut. The toxin produced causes feeding to grind to a halt when it attacks the stomach lining of the insect and begins breaking down the gut wall, allowing stomach enzymes and other bacteria into the insect's body. The combination of starvation and stomach tissue damage usually leads to death of the infected host within 2–4 days. Once dead, the bacteria are released into the general environment, enter a spore stage, and persist for protracted periods of up to 2–3 years until consumed by another vulnerable insect host.

A variety of strains of *Bt* have proven useful as an insecticide for a variety of crops. Cotton engineered with *Bt* genes utilize the *kurstaki* strain described above because this strain is highly toxic to the larvae of bollworms and budworms. Many larvae can be killed with very low toxoid doses because the young of these worms tend to feed rapidly, thereby ingesting more bacterial spores and increasing their body burden of bacteria. But, even then, as one ad in the *Delta Farm Press Journal* (30 Jan. 1998, pp. 2–3) admits, the level of protection claimed by Monsanto is incomplete or "subthreshold," keeping "worm damage at a lower level."

Too Much of a Good Thing?

Historically, *Bt* has been a boon to agriculture because it provided a self-limiting, organically approved, non-chemical alternative for insect control. As such it has proven to be a vital tool for organic farms. But massive release of this organism or impregnating plants with its genes on a large scale potentially threatens *Bt*'s ultimate usefulness. In the face of an abundance of the toxin, natural selection for resistant pest species may dampen its utility.

For cotton, the emergence of *Bt* resistant strains of budworms and bollworms poses a real risk once this historically effective pest control agent loses its effectiveness. Resistance arises when organisms are selected that have the ability to detoxify the *Bt* poison or prevent its uptake.

Such resistance must be seen against the background of pesticide resistance generally. Insect resistance to pesticides is a major problem in modern agriculture. Over 500 arthropod species are resistant to at least one pesticide.[4] One of the targeted pests for cotton crops is *Heliothis virenscens* or the tobacco budworm. This insect has evolved to be resistant to almost all registered conventional insecticides, creating a strong incentive for farmers to use plants containing the *Bt* toxin.[5] Fortunately, this budworm has proven to be highly susceptible to *Bt.*[6] A related cotton bollworm, *Helicoverpa zea*, is less sensitive to the toxin so that different strategies for using *Bt* for its control have had to be developed. When whole fields are planted with toxoid-containing transgenic crops, it is the *Helicoverpa* bollworm which can be expected to become resistant first.

Guidelines

In an attempt to postpone the inevitable loss of *Bt* susceptible hosts, the Environmental Protection Agency (EPA) has issued planting guidelines for transgenic crops. These guidelines are intended to reduce the likelihood of emergent, resistant insects by creating artificial niches for the survival of susceptible hosts. With the advent of mass plantings of *Bt* corn and the expansion of *Bt* cotton plantings in 1997, many of the natural refuges where susceptible larvae thrive will disappear. Ironically, the more product transgenic growers use, the more rapidly they will select for resistance in insects. By setting aside refuges, *Bt*-free traditional varieties in fields containing transgenic seeds, the EPA believes insects will not be as likely to form resistance. Leaving sections of acreage free of *Bt* containing plants reduces the selection pressure for resistance. But will this theoretical construct work?

To provide a specific example: if a farmer purchases seeds containing Monsanto's *Bt* gene technology, he currently has two options for planting. One option is to plant 96 percent of his fields with the *Bt* seeds, leaving 4 percent that can neither be treated with external conventional insecticides, nor contain any transgenic *Bt* plants. This strategy, in effect, gives 4 percent of his crops to the bugs. The second option is to plant 80 percent of the crop with the *Bt*-containing seeds, leaving 20 percent planted with traditional, non-engineered varieties which the farmer can treat with conventional insecticides. We note that both of these strategies ensure that *Bt*-containing and non-*Bt*-containing crops will be harvested together, making separation of chemical or *Bt* toxoid free crop virtually impossible.

The EPA believes that creating refugia of these sizes is sufficient to counteract the likelihood of creating insect resistance. The evidence from the field is not so reassuring. A recently published study predicts the 4 percent refuge mandate by the EPA will virtually ensure *Heliothis virescens* will be resistant to *Bt* within 3–4 years.[7] But the larger refuges have not yet been defined and no workable strategy for protecting the ecosystem from *Bt*-resistant pests has been promulgated.

From everything we have learned, *Bt* may become ineffectual for farming and organic gardening, generally. *Bt*'s survival as a viable pest control methodology for the small farmer using integrated pest management strategies is thus threatened by the intentional mass introduction of the *Bt* genes into plant tissues. The belated recognition of this likelihood has led the major manufacturers of the *Bt* gene, giants like Monsanto, Ciba-Geigy, and Novartis, to recommend the intentional creation of much larger insect refuges, on some 20–25% of every field planted with a *Bt* crop. In these refuges, the farmer is to plant non-genetically and hence "vulnerable" crops, giving the *Bt*-susceptible insects a place to breed and hopefully avoid the otherwise inevitable resistance to *Bt* toxin.

This strategy makes sense to an evolutionary biologist, but what will the farmer think of intentionally putting a portion of his valuable crop in harm's way to ensure that he has pests for next year? Certainly the logic of this enterprise will seem absurd to most farmers—why not let my neighbor put aside a portion of his crop for destruction? Then I can maximize my yield and not have to sacrifice my field for the good of the insects. This unfortunate realization is what Garrett Hardin described in 1969 as the Tragedy of the Commons, a circumstance historically brought on by the practice of sharing a common pasture in each village. Villagers who are asked to put only one or two of their sheep on the commons soon learn that if they (and only they) deviate from the public rule, they will benefit. In time, of course, everyone behaves selfishly, and the commons are destroyed. This, we are afraid, is one of the predictable consequences of the *Bt* revolution: Insect resistance will become rampant, and with it will come the loss of effectiveness of an essential ingredient in the integrated pest management control system for insect pests that relies extensively on natural bacillus spores of *Bt* for its effectiveness.

This danger was brought home on April 14, 1997. A team of scientists writing in the Proceedings of the National Academy of Sciences reported that 1.5 out of 1000 moth larvae carry the gene to overcome *Bt* toxin, literally a thousand times more than had previously been expected.[8] Thus, the *Bt* system of protection is exquisitely vulnerable to evolutionary change. If, as expected, it is presently being undercut by genetic engineering, it will be a major blow to biological controls across the board. The consequences of selecting for resistance in insect larvae may be an accelerated evolutionary shift that could cause substantial ecological disruption.

Bt Performance So Far

A key question, of course, is the validity of our arguments. Are the fears of *Bt* resistant pests reasonable? Let's look at the evidence so far. *Bt*-containing Bollgard® Cotton was grown in the United States for the first time in 1996. In that year 2 million acres, an admittedly high first-year saturation of *Bt* cotton, was planted in United States soil. During 1996, Texas and Louisiana farmers planting *Bt* cotton experienced unusually large infestations of cotton bollworm. Monsanto attributed this disturbingly high rate of infestation to unusually hot weather bringing what it termed, "an usually high bollworm pressure."[9]

In the face of this unforeseen event, many critics of transgenic *Bt* became alarmed about insect resistance. Several questioned the adequacy of the refuge percentages afforded by the EPA plan.[10] Others argued that insufficient concentrations of *Bt* toxin reached the pollen. But few persons questioned the overall wisdom of this short-term transgenic blitz. We are concerned about the multiple levels of shortsightedness regarding the transgenic use of the *Bt* toxoid. Industry has little understanding of the amounts of *Bt* that are expressed throughout the various parts of a transgenic plant. The concentration of *Bt* needed for effective control as resistance escalates may not be attainable in all engineered plants.[11]

Monsanto Company appears nonplused by the setback, characterizing the doubts about *Bt* as "theoretical" concerns. According to Monsanto representatives, using Bollgard® Cotton in 1996 saved a quarter of a million acres from being sprayed with formulated chemical insecticides.[12] In their view, any crop losses were offset by the economic savings in pesticide costs. And in fact, Texas farmers saved on average $33 per acre by limiting their use of expensive pesticides.[13] However, many farmers do not seem to agree. A group of them have filed suit in Texas courts, alleging economic injury from crop losses and the purchase of a defective product.

Comment

On a broader scale, we remain unconvinced of the wisdom of mass plantings with single resistance genes, whether for herbicide tolerance or *Bt* toxoid. *Bt* is an invaluable adjunct to IPM and organic farming which effectively reduces reliance on chemically based insecticides. In the long run, we believe the overuse of *Bt* engineered crops will undercut the effectiveness of this vital biological control agent. We see economic forces driving us inexorably to such overuse, as incentives for reliance go with profit. Microbiologists understand so little about this particular group of bacteria, especially how their genes react and interact within their new hosts, that such widescale use will inevitably be experimental at best and irrevocably destructive to biological controls at worst.

We are especially concerned that the introduction of novel toxic genes may effect the environment in as yet unforeseen ways. Without any method of field testing to anticipate these second order effects, we remain hostage to the economic factors cited by Monsanto. At a minimum, we believe it advisable to trace the consequences of release of these organisms. Currently, the EPA and national crop boards lack critical information regarding where these new seeds are being grown. When approached, the National Cotton Council reported that they do not have the figures on how much transgenic *Bt* cotton is being planted.[14] When the National Corn Grower's Association was asked which farmers were using Yieldgard® Corn, a respondent said, "I do not have that information, and frankly, I could not tell you where to go to track that information down."[15] In short, no one seems to be responsible for keeping tabs on this vast experiment upon nature.

A Corporate End-Game?

The most cynical view of the *Bt* story is that the apparent neglect of strategies that might otherwise favor *Bt's* continued utility may be intentional. Why would any business enterprise potentially encourage obsolescence of its own product? We believe economic motives can explain at least some of the behavior that contributes to the passive destruction of *Bt*'s effectiveness. While our viewpoint is speculative, there are several features that render it plausible.

First, to recap, the widespread insertion of *Bt* genes for toxoid production into corn, cotton and other commodities greatly increases the selection pressure favoring the emergence of *Bt* resistant strains of coleopteran and lepidopteran larvae. Second, the lion's share of the patents for *Bt* resistance factors and the technology for getting the requisite genes into plants belongs almost entirely to a single company, Mycogen corporation, holder of over 50 different patents, with a much smaller share belonging to Ecogen.[16] Each of these companies is partly or wholly controlled by developers of engineered seeds containing the *Bt* technology. And only Mycogen appears to hold patents on viable alternative *thuringiensis* strains to *Bt.*

In 1986, shortly after the first *Bt* technologies began to be patented, Mycogen formed an alliance with DowElanco (now Dow Agrosciences), itself a joint venture between Dow Chemical Co. and Eli Lilly. Dow Agrosciences has increased its holdings in Mycogen to 63%. Monsanto Company purchased a 13% interest in a competing company called Ecogen in January 1996, acquiring the rights to Ecogen's 10,000 *Bt* strain library for in-plant use.[17]

Mycogen, in turn, like its competitors Monsanto and Novartis, has begun to buy up seed companies. According to analyst Russ Hoyle, the rapid explosion of these acquisitions and subsequent expansion of *Bt* engineered seed crops is necessary to put Mycogen in a competitive position with its giant competitors.[18] But Mycogen scientists have estimated that the effective shelf life of their *Bt* technologies is no more than about 10 years. At about that point in their use, resistant crops will have far outstripped *Bt* insecticidal utility. According to Hoyle, "Even with ongoing efforts to slow or manage the development of resistance to *Bt* proteins in pests and insects, eventual resistance is virtually certain."[17]

To us, it appears plausible the major actors in the *Bt* wars who hold the oldest patents on this technology would actually benefit by extracting their profits "early," during the years while *Bt* is effective and their patent is in force. By allowing resistance to supervene, companies like Mycogen and Ecogen which have new toxic producers in the pipeline could benefit from this outcome. Once their patent expires, they virtually ensure that future competitors will find their own generic *Bt* products ineffective. Even if this strategy is more accidental than intentional, we believe the misuse through overreliance on *Bt* will favor Mycogen and probably have the most serious evolutionary consequences of all biotechnology innovations.

Political Action

In September 1997, in conjunction with Greenpeace and nearly 30 other environmental groups, our organization filed a petition to halt the continued registration of *Bt* crops. In this petition, the Greenpeace consortium declared the continued use of *Bt* transgenics threatens sustainable agriculture and may cause significant human health and environmental problems. Our own view is that by essentially committing genetic sabotage of this natural, limiting protein, biotechnology companies may well ruin a miracle. As we noted, the clock is running out on patent protection, leaving the major actors with a decade left of exclusive control. If, during that same decade, resistance is permitted to emerge, the result will be destruction of *Bt*'s effectiveness.

Anticipating this outcome, companies like Monsanto have already made plans to create new genetic miracles, for instance the genes that code for the expression of cholesterol oxydase. But this gene's utility has nothing to do with agriculture—and the residue left behind by the *Bt* gene wars will be a field in which *Bt* resistant insects flourish, out of the reach of this critical biological control agent. In the end, it is the regulatory community which serves as the gatekeeper for biotechnology as a whole. How well it does its job will determine how well this new technology will be integrated into the agricultural community. To date, the history of this regulatory effort is not at all reassuring.

---

[1] See Russ Hoyle, "Mycogen's Caulder steps down but not out," *Nature Biotechnology* 15: 487, 1997.

[2] Monsanto Annual Report, St. Louis, Mo., 1996, p. 16.

[3] B. E. Tabashnik, "Seeking the root of insect resistance to transgenic plants," *Proceedings of the National Academy of Sciences* 94: 3488–3490, 1997.

[4] G. P. Georghiou and A. Lagunes-Tejeda, "The Occurrence of Resistance to Pesticides in Arthropods," *Food and Agriculture Organization,* Rome, 1991.

[5] S. H. Martin, G. W. Elzen, J. B. Graves, S. Micinski, B. R. Leonard and E. Burris, *Journal Econ. Entomol.* 88: 7986–7988, 1995.

[6] J. Fox, "*Bt* cotton infestations renew resistance concerns," *Nature Biotechnology,* Vol. 14; 1070, 1996.

[7] F. Gould, A. Anderson, A. Jones, *et al.* "Initial frequency of alleles for resistance to *Bacillus thuringiensis* toxins in field populations of *Heliothis virescens,*" *Proceedings of the National Academy of Sciences* 94: 3519–3523, 1997.

[8] Tabashnik, B. E., Liu, Y. B., Masson, L., Heckel, D. G. "One Gene in Diamondack Moth Confers Resistance to Four *Bacillus thuringiensis* toxins," *Proceedings of the National Academy of Sciences* 94: 1640–44, 1997.

[9] USEPA, OPPTS, Biopesticide and Pollution Prevention Division. "*Bacillus thuringiensis* subsp. *Kurstaki* insect control protein as produced by Bollgard® gene in cotton, EPA Reg. No. 524-478. Submission of information concerning the acreage of Bollgard® cotton planted in the U.S. and the results of Monsanto's insect resistance monitoring programs," 5 November, 1996.

[10] Weppelman, Roger, Letter to Office of Pesticide Programs, 22 July 1996.

[11] Estruch, *et al.* 1996.

[12] Fraley, 1996.

[13] Barton, Gary. Press Release 071896. 18 July, 1996.

[14] Ann Rona, National Cotton Council, 13 April, 1997.

[15] Ann Burn, National Corn Grower's Association, 12 April, 1997.

[16] QPAT-US Patent Database, www.qpat.com.

[17] Westergaard, Mitchell, Enviro-Tech 2000. September 1996.

[18] See R. Hoyle, "Mycogen's Caulder steps down but not out," *Nature Biotechnology* 15: 487, 1997.

6

Regulatory Review

Who's in Charge?

In principle, the enabling legislation of our most powerful regulatory bodies, the United States Department of Agriculture (USDA), the Food and Drug Administration (FDA), and the Environmental Protection Agency (EPA) places the agricultural biotechnology industry within their purview for governmental oversight. We can identify three good reasons for intensive regulation: First, biotechnology can disrupt ecological systems. For instance, transgenic plants contain novel genes which may migrate to unintended weedy species, a concern of the USDA. Secondly, new transgenic products may contain novel proteins with allergenic or toxic properties, a focus for the FDA. Third, the plants produced may contain higher concentrations of oversprayed pesticides, making new tolerances necessary, a domain of the EPA. Examples of each of these untoward events have in fact been reported. Yet, as we will document, each agency has ducked responsibility and regulatory review has been blunted. Why?

The reasons for this lack of discretionary control are complex. Governance by three separate agencies for engineered plants is not unlike the regulatory patchwork quilt for other non-engineered foods. The EPA is responsible for the permissible amount of pesticides and herbicides on foods and sets "safe" levels of any residues; the USDA is responsible for overseeing the actual plant, how it is grown and where it is planted. The FDA enforces the tolerances set by the

EPA to make sure that proper labeling is placed on the food products. Under the provision of the Federal Food, Drug, and Cosmetic Act (FFDCA) of 1938 and the new Food Quality Protection Act (FQPA) of 1996, the EPA is responsible for establishing tolerance levels for the herbicides and pesticide residues on raw and processed foods.[1] Tolerance is defined as the amount of pesticide that may legally remain on the crop after harvesting. In turn, the FDA enforces the tolerance levels set by the EPA for all agricultural commodities and is responsible for product labeling. Food and meat processors who use transgenic products are responsible for making sure that they are in compliance with the FDA and EPA standards.

But there is at least one fly in the FDA's ointment. The FDA can perform an inspection at any time, much like the Internal Revenue Service can perform an audit on our taxes. But unlike your relationship to the IRS, a company subject to an FDA audit may preempt the regulatory agency by conducting a self-inspection. This new twist on the fox guarding the hen house has been met with great corporate enthusiasm. In recent years, the vulnerability of corporations to federal environmental audits has been greatly reduced by the promulgation of "environmental audit privilege laws." These laws provide for broad proprietary privileges as long as a company voluntarily divulges any alleged misconduct. In return, the company may demand the public be kept from seeing what errors or commissions put it out of compliance with the law in the first place.

In 1996, the three major regulatory agencies resolved their overlaps by putting the Animal and Plant Health Inspection Service (APHIS) under the USDA in charge of transgenic crops. APHIS regulates the planting, distribution, and harvesting of the actual plants that are transgenically altered. In view of the over 800 tests being conducted annually, APHIS's role in overseeing the permit process for conducting field tests would appear daunting. But APHIS has truncated its role by limiting the number of tests it must supervise. APHIS has in fact granted major corporations permission to field test many of their crops *without* prior regulatory review. Once deregulated, all oversight (with the exception of pesticide tolerances) ceases for transgenic crops, thus releasing them for full scale commercialization. Among the crops that are no longer regulated are Monsanto's Roundup Ready™ corn, AgrEvo herbicide tolerant soybeans, DeKalb herbicide tolerant corn, and DuPont's herbicide tolerant cotton.[2]

A Case Study: Regulation of Roundup Ready™ Soybeans

One explanation for the abdication of regulatory power and its passing from EPA to APHIS is that the EPA does not focus on the plant itself but instead on the introduction of new plant pesticides that are likely to result in novel environmental or human exposure.[3] In a technical sense, we can understand the EPA's abdication of control. When agbiotech companies introduce an herbicide-tolerant crop, they are not introducing "new" herbicides, they are introducing a vehicle for broader usage of an already approved herbicide. But the EPA may be overlooking two critical factors: Some gene products, like those for bromoxynil resistance, *change the herbicide into a new toxic form*, and transgenic plants often invite entirely different, usually greater, amounts of herbicide use than do traditional varieties.

Notwithstanding, companies like Monsanto have asked the EPA to increase the established allowable and "safe" tolerances for the herbicides to which they have engineered resistance. Because the residue levels would increase as the herbicide is directly applied onto existing plants, a modest increase in tolerance levels may be essential in order to introduce a given product line. The pressure on the EPA to permit *more* herbicide and higher tolerances for residues can be intense. By putting this pressure on well in advance of marketing the transgenic product line, a company can avoid the impression that they are asking for a special crop-specific exemption.

Getting Roundup® Out of the Corral

In 1987, the EPA supported an increase in the tolerance level for glyphosate, the key ingredient to Monsanto's incipient Roundup Ready™ technology. At this time, the alleged rationale was predicated on the use of Roundup® as a "post-harvest aid" to kill plants like soybeans to make them available for earlier silage. The 1987 tolerance level for glyphosate was set at 6 parts per million. In the late 1980s, when the Roundup Ready™ product line went from the drawing board to field testing, Monsanto claimed glyphosate tolerance levels up to 20 parts per million were safe to human health. The EPA accepted these representations. The increased tolerances, vital to the successful use of Roundup Ready™ technology, were in place just as Monsanto's first transgenic crops were planted.

Our research leaves open the possibility that the tolerance changes were instrumental to an overall strategy to prepare the soil for Roundup Ready™ crops. The United States government requires the company producing a product like Roundup Ready™ soybeans to supply the studies showing that the product is safe. In the case of Roundup Ready™ soybeans, Monsanto timed and produced studies for the USDA which established the safety of the plant in the environment, as well as studies that were used to establish safe levels of the herbicide Roundup® at the critical juncture of their production schedule.

While none of the submitted studies we have seen are questionable from a scientific viewpoint, Monsanto has been unwilling to share critical research information on toxicity with us.[4] In spite of the fact that the number of such studies submitted to the United States government is undoubtedly large, we do know that essential data gaps exist, especially for endocrine studies. For instance, when we asked the EPA if they tested for glyphosate residues in meat, our contact informed us that they do not conduct tests and that they rely solely on the companies to submit test results. He told us that no glyphosate tolerance is set for meat, only in kidney and liver.[5]

Regulatory Largesse

In May 1992, the FDA determined unless substantial, qualitative differences exist, all foods derived from new plant varieties produced by genetic engineering would be regulated no differently from foods created by conventional means.[6] In effect, the FDA has declared a state of equivalency between food products processed with genetically altered crops and those made from non-genetically altered crops. As a result, bioengineered foods can be found on grocery store shelves intermingled with non-engineered ones. But no distinguishing markings, notices, or labeling are to be found to allow the consumer to tell them apart. No labels alert the consumer to any potential health effects from eating foods with novel genetic material, including possible allergens, or potentially greater herbicide levels from overspray operations. In the FDA's view, since no significant difference exists between those transgenic and normally cultivated foods, no special notification is needed. Such a lack of discrimination invites the public to consider genetically altered foods as safe as non-altered products. We believe this conclusion is premature.

The FDA's position appears to be predicated on minimizing companies' concerns about problems in costly handling that might stem from labeling. If a labeling rule were enforced, the product's cost would likely increase because transgenic and non-transgenic products would have to be kept separate at all levels of processing. Industry rejects this option by arguing it would be impossible to separate cleanly engineered product from non-engineered product throughout the course of manufacture. Of course this state of affairs arises precisely because virtually all crop harvest combines intentionally mix engineered and non-engineered product in their silos. As proof that such obfuscation is unnecessary, consider Central Soya, one of the nation's biggest soybean processors. Central Soya currently separates transgenic and non-transgenic soybeans into different silos with no apparent difficulty.[7]

Regulatory Consequences: What Could It All Mean?

As a result of this growing laissez faire attitude of regulatory agencies, transgenic crops are being uncritically integrated into the American food economy. No comprehensive statute addressing the environmental risks of agriculturally engineered plants is presently in place, leaving a framework which only regulates these plants under the standard statutes governing any plant pest, food or pesticide issue.[8] Existing regulations also fail to factor in the full commercial scale use of transgenic crops, assuming that only small acreages will be planted. For soybeans, this is already a myth, as upwards of 20 *million* acres were planted in 1998. But any potential long-term effects to human health of these and other residues is almost certainly going to increase with the expansion of genetically engineered foods. This problem is compounded by extending exemptions to preempt any new reviews of novel transgenic crops coming into the pipeline.

One consequence of deregulation is that the USDA is relying on extrapolations from old scientific literature to assess the risks of these crops with little or no incorporation of new data.[9] The data in these older studies is typically extracted from field tests covering areas of only a few acres. This data is then used to support the safety of releasing the crops for commercialization in the millions of acres. As of late 1997, more than two dozen crops so evaluated were no longer regulated and have been commercially released. Existing field trials also preempt follow-up studies intended to learn of any long term effects of these crops on the environment. No protocol is in place to handle an accident that may arise from the widespread release of engineered organisms or their unintended spread to neighboring fields or closely related weedy species. Such a "head-in-the-sand" posture is alarming, given the propensity of transgenic pollen to move far off site and the existence of at least one report of transfer of engineered genes to weedy crops, points to be discussed below.

A Case Study: The Roundup Ready™ Field Trials and Commercial Release

Monsanto's Roundup Ready™ crops exemplify this pattern of early intensive review followed by a relaxation of oversight. The critical juncture for the successful marketing of Roundup Ready™ products occurred on December 6, 1993, when the Animal and Plant Health and Inspection Service (APHIS) acknowledged receipt of a petition from Monsanto to exempt its glyphosate tolerant soybean from further regulatory control. In this petition, Monsanto argued that its newly engineered soybean line, identified as 40-3-2, posed no novel risks to plants or humans; that it added no "plant pest risk" to the environment generally; and that it was extremely unlikely to be able to confer any selective advantage to other plants (weedy or not) that might lead to ecological disturbance or the emergence of a newly dominant plant type.[10]

In its petition, Monsanto expanded these three assertions to a total of seven claims, none of which spoke directly to concerns that the newly introduced plants might pose a novel risk of weediness or other potentially damaging environmental impacts. Monsanto did not address these issues and instead made commercial arguments for the economic value of their product. Among their assertions for transgenic crop benefits were:

1) the prospect of giving the farmer the option of using a broad spectrum herbicide;
2) the virtue of using an "environmentally sound" [sic] herbicide;
3) permitting the use of a chemical weed control in-season (i.e., while the soybeans are growing);
4) being able to treat field weeds on an "as needed" basis;
5) offering less reliance on pre-emergence herbicide use (because more glyphosate can be used later);
6) being useful within no-tillage systems (avoiding having to plow in crops after the growing season); and,
7) providing a "cost-effective" weed control.

None of these points was documented with supporting material. In our view, these assertions served to deflect attention away from what should have been the main issue for APHIS: Will glyphosate resistant crops pose novel ecological risks? Instead of examining all of the facets of this issue, APHIS asked only if glyphosate tolerant soybeans posed risks as "pests" themselves or through gene transfer to weedy relatives. Of course Monsanto gave assurances this could not happen. In the end, having reviewed only limited arguments for the environmental safety of soybeans, APHIS approved Monsanto's application.

We question the objectivity of this decision. In responding to our concerns about long-term safety, Mr. Arnold Foudin, Deputy Director of the Biotechnology and Scientific Services Division of APHIS/USDA declared Roundup Ready™ technology to be a "boon" to farmers and agriculture generally. He also dismissed our concerns about higher residue levels from the increased reliance of glyphosate generated by Roundup Ready™ crops. He stated, "Weeds are the number one pest for farmers and Roundup® is being developed into this technology because it works, is environmentally friendly, and becomes a free lunch for the microbes."[11] While Foudin may have intended to mean microorganisms feed on the Roundup® that binds to the soil particles, he ignored the reality that glyphosate can inhibit certain soil microorganisms and fungi, affect intestinal bacteria and disrupt the soil's micro-ecosystem. He also missed a key point about microbial metabolism: some bacteria can convert Roundup® into more toxic substances, a point we address in the conclusion.

It is argued that because the novel gene sequence introduced into its soybeans coded for an enzyme used only by plants, it posed no risk whatsoever for any non-plant species.[12] This, of course, is only true in degree. The enzyme glyphosate poisons, known as 5-enolpyruvyl-shikimate-3-phosphate synthase or EPSPS for short, not only assists plants in making certain amino acids that are used in photosynthesis, but also determines indirectly how much of these critical amino acids will be left in the edible portions of the plant. The amino acids produced are so-called aromatic ones (i.e., those having a benzene ring) such as tryptophan and are essential amino acids in many other species including mammals, birds and fish. Animals typically acquire these aromatic amino acids ready-formed and cannot synthesize them from precursors. Hence, a genetically engineered plant may have altered amounts of critical amino acids in its edible portion.

Monsanto's conclusion that because this aromatic amino acid pathway is exclusively present in plants means that glyphosate "Is only toxic to plants but not other living species, including mammals"[13] is, in our opinion, simply wrong. While only moderately toxic, glyphosate clearly is toxic for species other than plants, albeit at relatively high doses. As we reviewed in Chapter Four, glyphosate can cause liver toxicity at high doses and may have gene damaging potential in some plant species.

More critically, while glyphosate may be a relatively benign herbicide when tested on individual species in tissue culture or outside an organism's natural environment, its aggregate impact on the mix of soil microorganisms that survive repeated applications remains a serious unknown. Additionally, genetic risks must be considered. In theory, only a single gene remnant of the original genetic material from the vector chosen by Monsanto to put their engineered product into soybeans is present in the soybean line or strain in question. Of concern to the USDA is the prospect that other genes from the original vector, the potentially pathogenic *Agrobacterium* survived the gene insertion process.[14]

Also left unanswered is the mystery of where the new gene(s) actually end up in altered soybean genomes. While the end result, a glyphosate resistant plant, is unquestioned, it was isolated from among many gene-engineered plants by random human selection for glyphosate resistance, not on the basis of any scientifically sound gene screen. Only after plants with the desired physical characteristics were isolated did Monsanto know it had successfully engineered a product. Further after-the-fact testing did reveal that the novel gene was integrated into the soybean genome at some indeterminate locus, making it possible for Monsanto to cross the 40-3-2 variant with traditional varieties, mate them to show Mendelian inheritance and still retain the novel attribute of glyphosate resistance in a portion of the progeny.

Because of this blind insertion system, the possibility that the newly introduced gene destabilizes the genome in some way remains a critical possibility. Proving where each new gene goes is a daunting but essential project. For instance, on some test plots treated with Roundup®, a few 40-3-2 plants appear to grow taller than their non-engineered neighbors, a completely unanticipated consequence of the genetic manipulation of the EPSP genetic pathway. Could these plants contain the novel gene in a unique chromosomal locus? Would not scientists want to know? It is also plausible that the ability to make more of the critical amino acids at risk in glyphosate sprayed plants only confers a survival advantage to the plant under herbicide-saturated treatment conditions, but leaves it vulnerable under nonspecific forms of environmental stress.

Other data we have seen on side by side yield comparisons appear to show some Roundup Ready™ plants to be growth deficient compared to their genetically normal counterpart. Specifically, in the Roundup Ready™ soybeans use guide prepared for the Delta/Mid-South (©1996 Monsanto Company #004-95-49A), hand-tilled soybeans yielded 38.2 bushels per acre while all eight test plots had lower yields (ranging 36-38 bushels/acre). Yield reduction data such as these raise the possibility that the new gene confers a selective *dis*advantage to the plant for growth in the absence of glyphosate application. If true, data such as these cast doubt on Monsanto's principal point that their genetic engineering is both botanically and environmentally neutral. To ascertain if Roundup Ready™ plants do have comparable yields to conventional varieties, we scrutinized the University of Arkansas' data (see Figure 5).

Figure 5. Comparative Yields: Roundup Ready™ vs. Conventional Soybeans, Arkansas, 1996

SEED COMPANY	RR VARIETY	YIELD	CONVENTIONAL VARIETY	YIELD	DIFFERENCE
Hartz	4994 RR	62	4994	63	-1
		43		47	-4
	5764 RR	54	5050 †	60	-6
		54		56	-2
		38		46	-8
		37		41	-4
		47		46	+1
		47		48	-1
		20		31*	-11
		20		25	-5
		37		59*	-22
		41		56	-15
		55		62	-7
		44		55	-11
		31		36	-5
		39		46*	-7
		76		36	-10
	5545 RR	58	5545	59	-1
		55		60	-5
	54989 RR	56	517462	52	+4
		56		52	+4
		51		55	-4
Delta King	5961	59	5850 †	64*	-5
		41		52*	-9
		53		51	+2
		25*		26*	-1
		57*		58*	-1
		56		59	-3
		31		47*	-16
		51*		49*	+2

Graphics by Platt & Company

Terral	5666 RR	57	5797†	62	-5
		44*		40	+4
		45		48	-3
		24		23	+1
		48		52*	-4
		54		57	-3
		44		39	+5
		42		48*	-6
	AVERAGE	44.79		49.13	-4.26

$t = 4.34$; $p < 0.001$.

All Values in annual bushels per acre averaged over 2 years.

† Matched by common maturity date, growing conditions, disease resistance and irrigation.

* Equal to or greater than the maximum yield of the varietal grown in the same region.

Source: L. O. Ashlock *et al.*, Cooperative Extension Service, Soybean Update, University of Arkansas, February 1997.

Reduced Yields

In one study, we compared the actual performance of Roundup Ready™ soybean varietals with those of genetically identical or closely identical conventional varietals grown under the same conditions. Our data source was the 1996 figures assembled by the Arkansas Cooperative Extension Service, headed by Dr. Lanny O. Ashlock. We matched a Roundup Ready™ variety to its nearest conventional type and compared the number of bushels/acre harvested at the end of the 1995 and 1996 growing seasons for each type. In 30 out of the 38 comparisons, the conventional variety outperformed the Roundup Ready™ variety. The likelihood of such an outcome occurring by chance is less than 1 in a hundred ($Chi^2 = 6.95$, df 37). Overall, the yield per acre was down an average of 4.34 bushels for Roundup Ready™ varieties, a statistically significant loss of just under 10% compared to conventional types. In only 4/38 instances did the Roundup Ready™ crop approximate the yields of the highest yielding conventional soybean varietal type grain in the region. (See Figure 5 on pp. 82-83).

If transgenic crops "lose" some of their hybrid vigor as a cost of gene insertion, a prediction made by theoretical geneticists, any reduction in yield jeopardizes the success of the entire Monsanto project. When reached at Monsanto's corporate headquarters in St. Louis, Missouri, Dr. Eric Johnson, Crop Protection Division, decried the "premature" publication of data suggesting reduced yields and questioned its accuracy, stating that Monsanto's own studies of "isogenic" lines of plants (i.e., those differing at just one gene from each other) show no difference in yield. Such discrepancies leave unanswered the major question of whether the transgenically created crops widely used in cultivation are strictly equivalent to Monsanto's isogenic lines. If not, will they provide the true benefits promised by the corporation, and at what cost to the consumer and farmer? To answer this question, we took a long look at Monsanto's and our own government's assertion that biotechnology would feed the world.

In a June 1997 speech to the International Grains Council in London, England, Dan Glickman, Secretary of the USDA, argued the world has no choice but to accept the American position that transgenic crops are safe and essential to stave off world starvation.[17] According to Glickman, genetically engineered crops are the "second revolution in food production." He declared, "Biotechnology holds out our greatest hope of dramatically increasing yields."

---

[1] K. MacKenthum and J. Bregman, *Environmental Regulations Handbook,* Lewis Publishers (London, 1992) p.17.

[2] APHIS/USDA, Biotechnology and Scientific Services. Crops No Longer Regulated by USDA. 6 August, 1997.

[3] Draft Proposal, Office of Pesticide Programs, U.S. EPA #OPP00343. 18 Dec., 1992.

[4] Lemon, Martin. Telephone interview. 2 April, 1997.

[5] Errico, Phil. Telephone interview. 24 March, 1997.

[6] Federal Register Volume 57, p. 22986-88. May 29, 1997.

[7] Hoots, Dan. Central Soya. Telephone interview. 24 March, 1997.

[8] Margaret Mellon and Jane Rissler, *The Ecological Risks of Engineered Crops,* Cambridge: MIT Press. 1996.

[9] *Ibid,* p. 123.

[10] United States Environmental Protection Agency, Office of Prevention, Pesticide and Toxic Substances. "Glyphosate in/on Genetically engineered Soybeans. Evaluation of Residue Data and Analytical Methodology, PP#4369/4H5701; ID#000524-00445," 29 June, 1995.

[11] Arnold Foudin, Deputy Director of APHIS/USDA Biotechnology Scientific Services. Telephone interview. 6 August, 1997.

[12] Matsumura, Fumio, University of California, Davis. Telephone interview. 27 March, 1997.

[13] Diane B. Re, Monsanto Company, Petition for Determination of Nonregulated Status to M. A. Lidsky, APHIS, USDA, 14 September, 1993.

[14] Federal Register Volume 58, No. 232. 6 December, 1993

[15] Re, Diane, Regulatory Affairs Manager, Monsanto. Memo to Michael Lidsky, Deputy Director BBEP, APHIS, USDA. "Petition for Determination of Nonregulated Status: Soybeans with a Roundup Ready™ Gene," 19 November, 1993.

[16] Monsanto Company, *Introducing Roundup Ready™ Soybeans*, 1995.

[17] Dan Glickman, Remarks of USDA, International Grains Council, London, England, 19 June 1997.

7

Transgenic Products and Food Production: The World Food Problem

"The parched eviscerate soil gapes at the vanity of toil, laughs without mirth. This is the death of earth"

—T. S. Elliot

The advent of the biotechnology agricultural revolution undoubtedly heralds a transformation of the world's food supply. But the nature and extent of these changes remain largely speculative. Will they simply increase corporate profits or will they actually increase the world's food supply? Several major biotechnology companies have rationalized their agricultural programs by asserting their products will feed the world. Representatives of Ceregen, Monsanto Company's agricultural biotechnology unit, declared in a 1996 statement, "By the middle of the 21st century, the world's population will have doubled. How can the earth and its already strained resources sustain these additional billions without going ecologically bankrupt? Ceregen's answer: smarter, genetically engineered crops—crops that require less pesticide, use less water, yield more bushels per acre and pack more nutrition."[1] How much of this declaration should we accept?

First, the population pressure on our food supply is likely to be exaggerated. In contrast to Ceregen's projections, by the year 2030, the United Nations predicts 7.1 billion people will populate our planet, up a modest 1.2 billion from 1997 levels—not a doubling. Current trends make it extremely unlikely that another 5 billion people will be produced between 2030 and 2050. Even at these projected levels, it is appropriate to ask "will there be enough food"? According to Worldwatch Institute President Lester Brown, "the Green Revolution has to happen all over again" to keep pace with even this more modest projection.[2] For many in the biotechnology sector, transgenic crops are the key to this new revolution. But is the present generation of genetically engineered crops really the answer to world hunger? Let us reexamine how genetic technology is being applied.

The harsh reality is the vast majority of staple grain crops harvested each year is used to feed cattle and other livestock, not people. While it is technically true that a fraction of grain is converted to beef or pork, for most of the poorest people in the world, meat represents an unobtainable luxury. Corn and soybeans are planted on 140 million U.S. acres. The U.S. alone uses 2.9 *billion* bushels of corn to feed 39.6 million calves and 22 million cattle.[3] From one bushel of soybeans equaling 60 pounds, approximately 40 pounds of soy meal are set aside to feed livestock.[4] The use of feed supplements on this scale helps to support the prices of soybean, corn and other feed crops. The U.S. beef industry continues to generate close to $40 billion per year, and leaves less than 10% of planted forage crops to feed people in the U.S. and elsewhere.

Chemical companies also benefit greatly from having land farmed to feed animals, since animal feed carries far less stringent pesticide tolerances than does feed intended for human consumption. The net result of using transgenic crops to feed animals is that more chemical can be used. Companies like Monsanto also improve their profit margin by restricting farmers to specific herbicides that can only be used on their own herbicide tolerant seed while arguing that the eventual product (for example, whole soybeans plus the hulls) will be fed almost exclusively to animals.

A further problem is the integration of herbicide tolerance and cropping practices. According to Monsanto, one way herbicide use promises greater profitability is by reducing reliance on tillage. Tilling the soil less is considered an environmentally benevolent action which decreases the loss of soil from wind and rain erosion and protects precious microhabitats. But less tilling leads to more perennial weeds, and perennial weeds require more herbicide. In the last decade herbicide usage has risen by 250 million pounds.[5]

Overview

The United Nations contends poor resource management is to blame for much of the world's food woes. Improper and overuse of pesticides and poor farming methods are taking a toll on food production with 300 million hectares of farmland worldwide severely degraded and 1.2 million hectares showing moderate fertility loss.[6] By this token, United States based companies such as Monsanto, AgrEvo and Dupont are perpetuating further chemical abuse and potential farmland degradation.

Presently, two-thirds of all transgenic food crops are being engineered for herbicide tolerance.[7] According to Rebecca Goldburg of the Environmental Defense Fund, 10.5 million taxpayer dollars are being spent on the development of herbicide tolerant crops. U.S. governmental agencies are more than silently supporting this movement. Without a critical reevaluation, we are concerned the movement towards universal plantings of herbicide resistant transgenic crops will create a political reality. This reality will commit a large proportion of our agricultural resources to a transgenic economy. In our view, this dilemma is aggravated by corporate policies that are fueled by self-interest and not global altruism.

The principal strategy of many agbiotech companies is to ensure that its agricultural chemicals match its engineered seeds and any future germ plasm put into its engineered crops. By creating selective demand, companies like Monsanto and DuPont get a "lock" on a future commodity. Financial analysts and corporate planners, who chart the stock markets, closely watch these chemical companies and their product lines. Not surprisingly, virtually none of the newly engineered seeds are designed to meet world demand for increased food supplies. Instead, they are designed to meet existing and projected markets *only* in those countries with an ability to pay for the expensive infrastructure needed to support transgenic crops. Economists, along with executives of companies that make agricultural herbicides, know that as such plantings increase, so will the production of herbicides.[8]

Even by Monsanto's standards, few if any of the engineered crops have achieved significant yield increases.[9] Given this admission, we reiterate our view that any further increases in crop yields in modern food crops will almost certainly come from building on traditional breeding methods—not from transgenics. For instance, focused agricultural practices, including selective applications of nitrogen fertilizer with site specific fertilizer assessments almost certainly can make up any shortfall in the yields needed to keep up with population growth.

We admit it would be valuable to increase the kind and quality of plant proteins to reduce the impact of continued population pressure. In theory, transgenic interventions can increase isolated proteins. But few such efforts have been successful (or even made) to increase the nutritional value of the end product. Yield increases would also be invaluable. But still fewer, if any, single gene innovations exist that consistently increase overall yields. By conventionally breeding rice strains (*Oryza sativa*) with a wild, weedy type maintained in seed banks (*Oryza rufipogon*), Chinese researchers promise to increase yields by 20-40%. In contrast, *none* of the transgenic varieties produced to date even approach these figures. Although modest increases of 6-7% have been reported by Monsanto for its Roundup Ready™ soybeans, as we showed, most yield data show a loss over conventional crops. Our view is that non-transgenic means promise greater benefit. A recent review highlighted the value of "high precision farming" techniques coupled with conventional interbreeding of wild and progenitor strains and endorsed these non-transgenic approaches as the best bet to increase yields.[10]

Overall, most of the emphasis in farming to meet the world food needs is on creating the optimal environments to realize the yield potentials *already* present in current varieties. Once this is achieved—even if present distribution problems remain—many economists believe we will be able to keep pace with increasing population growth. Even if we only make modest gains in the status quo, International Food Policy Research Institute (IFPRI) economist Mark Rosegrant has stated, "We can, I think, feed everyone, even if we will continue to have problems distributing it equitably."[11]

Solving the World Food Problem

This view is clearly at odds with corporate pronouncements. In discussions with friends and colleagues, Monsanto's CEO and president, Robert Shapiro, has widely touted Roundup Ready™ technology as a means of solving the world's food problems.[12] Monsanto's main transgenic researchers have likewise touted the power of biotechnology to provide a revolutionary "fix" to some of the agricultural sector's most intractable problems. According to early declarations of Monsanto researchers, biotechnological innovation in plant biology offered an unprecedented opportunity to increase the utility of agricultural methods. In 1989, these scientists identified three specific objectives for the food industry and the specialty chemical industries of which Monsanto is a part: enhancing starch production; enhancing oil yields or changing their composition; and creating plants that express proteins with nutritionally balanced amino acid composition.[13]

To the best of our knowledge, eight years later *none* of these praiseworthy objectives is a visible part of Monsanto's corporate plan or production activities. While a corporation is always free to choose its strategic investments, a company of Monsanto's size operates on the "opportunity principle." Once resources for a given enterprise are committed on a mass scale, little or no opportunity for shifting towards a different venture remains. In the case of transgenic technology, this has meant Monsanto has chosen food products that have augmented their chemical business rather than increase the supply of food in the global economy.

Food for All?

Irrespective of transgenic crop's role in expanding world food supplies, consumers will be ingesting greater amounts of bioengineered foods in their diets over the next three decades. Foods and additives derived from transgenic crops have already permeated much of our market produce with nary a decimal point increase in nutritional value. Oils, emulsifiers and additives used in processed foods have increasingly consisted of transgenic plant byproducts but virtually none contain engineered improvements in nutritional value. While transgenic soybean, corn and cottonseed oils permeate the processed, ready-to-eat foods common to the Western diet, most have not yet been engineered for improved quality. To our knowledge, only one new oil is being transgenically designed to have a better nutritional balance of fats and oleic acids.

In 1996, 80% of all oil consumed in the United States was soybean oil[14] and an estimated 15% of that was transgenic in origin. Only 5% of all soybean meal is used for human consumption. Of this amount, the hulls are used in processed, high fiber bran breads, cereals and snacks. Soymilk is widely used in pediatric products. It is of concern because its estrogenic flavonoids (discussed below) can be found in infant formulas used to replace or supplement breast milk fed infants. In 1997, infant soy-based formula, using soy protein isolate powders, was used as a nutritional alternative for nearly 7% of all American infants.[15] In lactose intolerant babies and infants, all or most of infants' nutritional needs are being met by exposure to transgenic products whose nutritional value and possible side effects remain untested. While it is *assumed* that transgenic soy and non-transgenic soy products are equivalent to traditional milk sources, the possibility of amino acid differences coded by the EPSPS transgene, altered levels of estrogenic substances, or novel allergens remain unexamined. If any problems arise from transgenic crops, infants will likely be the first to show them.

Soy lecithin, another product created from the soybean, is used as an emulsifier and stabilizing agent in chocolate, candies, and hundreds of other food products.[16] At Easter time in 1997, the undisclosed transgenic origin of much of this lecithin led to the Swiss government's decision to confiscate and destroy 500 tons of locally made chocolate.

The United States is also providing other countries with specialty soybeans to be used in traditional foods. Hartz™ Seed Company presently has farmers grow variety 922 strictly for export to Japan for use in natto, a fermented and cooked whole soybean used to top rice.[17] Hartz Seed Company would like to convert the Japanese to a transgenic variety. But Hartz has been unable to sell Japan genetically altered soybeans because of stiff consumer opposition, although they are negotiating with their buyers to expand their sales of Roundup Ready™ natto soybeans in the near future.

Clearly, the Japanese market is an attractive target for transgenics. In June of 1997, Hartz spent a week negotiating with Japanese importers discussing the projected 100% inclusion of Roundup Ready™ seed by the year 2000. Hartz™ hopes to capitalize on the fact that Japan is actively importing soybeans from the United States because they cost less than soybeans grown in their own country. It is a huge and potentially lucrative market. American soybeans sell in Japan for 2700 yen per 60 kg, whereas Japanese soybeans sell for 24,000 yen per 60 kg, making tofu, for example, twice as expensive if made from Japanese grown soybeans.[18]

Separating Engineered from Natural

As part of our research, we surveyed major U.S. food manufacturers by phone. Although few, if any, consumers are aware, products such as Crisco®, Kraft® salad dressings, Nestle® chocolate, Green Giant® harvest burgers, Parkay® margarine, Isomil® and ProSobee® formula for infants, and Wesson® vegetable oils all incorporate genetically altered soybeans.[19] Makers of McDonald's french fries, Fritos®, Doritos®, Tostitos® and Ruffles® chips have verified that they too are using oil from genetically engineered soybeans. An idea of their rationale for not disclosing this fact publicly can be gleaned from a comment from a representative of Frito-Lay®, who simply stated "Frito-Lay® does not discriminate [sic] against the genetically engineered beans."[20] Ross Labs, the maker of Isomil® soy-based formula and Mead Johnson, maker of ProSobee® infant formula, have also verified that they are "not discriminating" between conventional and transgenic soybeans. They will use transgenic products if such beans are sold to them.

Since both transgenic and "natural" soybeans are routinely mixed in grain silos (except for seed), such a position is tantamount to admitting use of genetically engineered crops. A further result of this commingling is that many other products also contain the byproducts from genetically engineered soybeans, including virtually any factory processed, non-organic oil based food product in the United States. Given the growing awareness—and in some instances general public concern about buying or consuming genetically engineered products—we find this state of affairs troubling.

Genetic Byproducts in Foodstuffs

Our research shows that in the United States, genetically engineered grains and legumes are also routinely mixed with non-engineered products when they are sold on the open market. As such commodities move into international trade, this practice creates difficulties for foreign consumers. The large grain traders such as Cargill and Archer Daniels Midland (ADM) buy and process mass quantities of grains and legumes, press them into oils and then sell the oil to major food processing companies. Because of the "non-discrimination" policies of these two companies,[21, 22] the buyers of Cargill and ADM products cannot know if they are buying grain including little, none or all genetically engineered material. This is a major problem since some foreign buyers are presently required to segregate the genetically engineered from the non-genetically engineered foods by law or practice. Because, to the best of our knowledge, no records are being kept of what is being placed into the silos at harvest time, a precise answer to the question, "Are you using genetically engineered seed?" is difficult, if not impossible to give. But a general answer is possible. As of 1997, it is a safe assumption that a growing portion of cotton, soy and corn oil is derived in part from genetically engineered crops because of the volume of altered crops that have been grown in the last few years. The volume is so great that biotechnological innovation is changing the face of agriculture itself. To fully appreciate the impact of transgenic crop production, we must examine the extent of its use.

---

[1] Monsanto Annual Report, 1996.

[2] Mann, Charles. "Reseeding the Green Revolution," *Science* 277: 1038–1043 (1997).

[3] J. Lawrence and D. Otto, "Economic importance of the United States cattle industry," Iowa State University: Ames, Iowa, 1994.

[4] Frank Flighter, edible oils consultant, U.S. Soybean Board, telephone interview, February 26, 1997.

[5] "Pesticides: after slip in 1995, production regained ground in 1996," *Chemical and Engineering News,* 23 June, 1997.

[6] "The United Nations convention on biological diversity: a constructive response to a global problem, Earth Summit +5; Special Session of the General Assembly to review and apprise the implementation of agenda," 21 June, 1997.

[7] USDA/APHIS biotechnology permits, Notices of Release of Transgenic Crops, January through March 1997.

[8] Labor Department and the Federal Reserve Board Calculations, *Chemical and Engineering News,* 31 March, 1997.

[9] Monsanto Roundup Ready™ pamphlet.

[10] Mann, Charles. "Reseeding the Green Revolution," *Science* 277:1038–1043 (1997).

[11] Rosegrant, Mark, IFPRI economist, cited in Mann, Charles, p.1043 (1997).

[12] Halpern, Charles. Personal communications. 19 August, 1997.

[13] Gasser, C. S., Fraley, R. T. "Genetically Engineering for Crop Improvement," *Science* 244: 1293–99 (1989).

[14] Frank Flighter, Edible Oils Consultant, Telephone interview, 26 February, 1997.

[15] Federal Register Notices, 93-148-1, Vol. 58;232, December 6, 1993, p. 60.

[16] G. A. Norman, "Soybean Physiology, Agronomy and Utilization," Academic Press: New York, 1978.

[17] Larry Spooner, Hartz™ Seed Company, Telephone interview, 16 July, 1997.

[18] Aschi Shimbun, "Japanese campaign to promote food security," *Japan Almanac,* 1997.

[19] Telephone interviews, February 20-21, 1997.

[20] Theresa Broadbent, Frito-Lay® Nutrition Specialist, Telephone interview, 3 April, 1997.

[21] Thrane, Lisa, Cargill Public Relations. Telephone interview. 24 March, 1997.

[22] Reed, Jack. Archer Daniels Midland, Public Affairs Representative. Telephone interview. 2 February, 1998.

8

Biotechnology's Impact on Agriculture

In the past, technological advances in agriculture have been made faster than the framework of regulation and public oversight have adjusted to them. We have seen this happen with over-intensive agriculture in the 1920s that culminated in the "dust bowl," and with the overreliance on pesticides like DDT and parathion for pest control. We have seen the aftermath of overuse of fertilizers, resulting in extensive contamination of aquifers and waterways with nitrates. And we have seen it with the blind export of Green Revolution technologies leading to food shortages in countries that could not afford the required irrigation and fertilizer.

We believe we are at a similar juncture with agricultural biotechnology today. Even as more and more crop releases are planned, we remain uncertain of the long-term consequences of the wholesale shift to herbicide-tolerant or *Bt* containing food crops. Researchers lack the basic motivation (although not the technology) to track where their genes go or how their chemical dependencies affect other organisms in the micro-ecosystem. We have no master plan to chart what will occur on a regional or planetary level as more and more cropland is converted to bioengineered crops. This lack of knowledge is disturbing and should give us pause.

We admit to a fundamental skepticism of the motives of the progenitors of this new technology. Many of the companies that once made environmentally unsafe chemicals like Agent Orange or PCBs have regrouped to form biological subsidies. Many have spun off new subdivisions to deal directly with biotechnology. Some like DuPont, Dow, and Monsanto have shifted into the development of genetically engineered products and formed new life sciences divisions. Typically, each company markets only the living product genetically programmed to intensify the use of its own specialized chemicals.

The fact that a chemical company has expanded into a life science biotechnology firm carries no moral weight of its own. But the word "biotechnology" has a certain euphemistic quality. "Bio" meaning life, gives the illusion that companies are producing ecologically safe, life-enhancing products. In its more literal translation, biotechnology means the application of science to life forms for commercial objectives. In essence, this anthropocentric science is necessarily neither ecologically safe nor harmful by invention, unless and until its objects of production enter the biosphere. Spread of some transgenes is virtually assured. Unfortunately, few of the seeds created by genetic engineering are non-propagative, and hence nonviable by design. Fertile transgenic pollen can and has escaped to contaminate weedy species.[1] Even this unfortunate side effect could have been offset by encoding the genes for male sterility or selfing genes that prevent fertilization. Without such built-in protections, gene flow between transgenic and native species remains a disturbing possibility.

The full panoply of effects from the release of millions of genetically engineered crop plants are presently uncertain. At one level, they may have no greater effect than do conventional non-engineered crops. At another, they may produce subtle and lasting change. Years of human ingestion of genetically engineered foods, or of agricultural ecosystems subjected to chronic exposure to chemical residues from herbicide treatments may lead to major health or environmental changes. Certainly the ecosystems next to transgenic food crops will experience some lasting effects. But the full gamut of secondary effects remains unknown.

Species Composition

A major concern is that the integrity of plant species will be compromised as engineered crop acreage increases and pollen mediated gene flow swamps and suppresses rare native plants that are congeners for transgenic crops. Biodiversity may also be reduced by heavy herbicide reliance. As remote as the present prospect for gene flow seems, herbicide targeted weedy species may pick up some transgenes and evolve more quickly than now anticipated. Many of these and related second order consequences of transgenic agriculture are intricately linked to ecosystem diversity.

Biodiversity

Agricultural biotechnology threatens to decrease the number of crop plant varieties currently grown by substituting a few varieties for the many now in commerce. Take corn for example. In order to have a patent on the technology for transgenic corn, biotechnology companies have to prove that they have constructed and will maintain a *uniform product.* Such a requirement guarantees the genetic distance present among genetically engineered seeds will be kept extremely narrow. Any resulting progeny will be similar, if not identical, to one another.

Concern that genetic diversity will be sharply curtailed, even among traditional cultivars, is underscored by a series of events culminating in Pioneer Hi-Bred International Seed Company's decision not to offer transgenic seed corn. Beginning in 1995, Monsanto and Pioneer began negotiations for Pioneer to carry Monsanto's Roundup Ready™ corn. But on November 13, 1997, after two years of talks, Pioneer abruptly pulled out of the deal, stating in a letter to its customers that the contract's terms "could significantly limit the number of traits, genes, and technologies" among the types of corn Pioneer intended to market.[2] This decision marks the first time a major seed company has gone public with its concerns about the restrictive covenants and conditions of Monsanto's contracts. More to the point, Pioneer's courageous decision underscores the value seed companies place in maintaining a diverse stock of seed germ plasm, a vital necessity for ensuring crop characteristics matched to shifting growing conditions and climatological change.

Pioneer's press release declared "we concluded that Monsanto wanted to ultimately determine what additional traits could be included in those [transgenic] products and the price to be charged to the farmer for those traits."[3] In response, Monsanto stated it was committed to making all its crops widely available and "regrets" Pioneer's news statement because it "put an inappropriate spin" on the negotiations.[4] Monsanto appears to have missed the point: it is not how many seed companies to whom they have successfully offered their technology, but how restrictive their offerings have been.

If the plants being engineered were confined to the United States, the issue of decreasing biodiversity might not be a major one. This is because the United States is not a center of biodiversity for the crops that have been commercially engineered. But most of the transgenic crops are slated for worldwide distribution in countries such as China, Thailand, India, Brazil and South Africa. These countries are in high species diversity zones, increasing the risk for biodiversity disruption. In these countries the proximity of weedy species closely related to the engineered type may permit genetic transfer of the novel engineered property from transgenic to native species.

Agricultural diversity was traditionally maintained by farmers in microecosystems sharing the best adapted crops that came from each year's planting. For thousands of years, farmers exchanged seeds allowing them to maintain a dynamic portfolio. Often the resulting broad-based gene pool proved essential to protect their fields from blights or other depredation. In many indigenous cultures farmers inherently knew the value of keeping their fields diverse. If a blight or rot attacked one variety, another would likely be immune, saving their fields from total loss. Historically, such variation and diversity have assured protection of food supplies. Why then are we moving away from this traditional goal and towards transgenic monocultures? The answer turns on economic factors.

Commercialization

For a properly equipped farming operation, genetically altered seeds cost a premium and rely on expensive herbicides, but still save tens of dollars an acre in production costs. Transgenic seed is designed by the manufacturer to favor economies of scale, saving the large-volume farmer more money as increased acreage is devoted to engineered crops. Producers of genetically engineered crops know these facts and cater to large commercial farms. Such a tendency will exaggerate an already disturbing trend. The small noncommercial farms are being lost progressively to large commercially intensive farms on an ongoing basis. In the United States, for example, 73% of the farms reported sales of less than $50,000 and contributed only 10% of all farm income.[5] In contrast, the 2.2% of farms with annual sales of $500,000 or more accounted for almost 40% of farm income. The dominance of large scale operations leads to greater reliance on crop uniformity. And crop uniformity means reliance on fewer and fewer germ plasm lines of commodity crops like corn, wheat, or soybeans.

Domestication of Seeds

As we continue to domesticate our seeds, the genetic diversity afforded by normally diverse genomes will be lost in the name of uniformity. This uniformity goes against the grain of evolutionary forces which have selected for plant diversity over tens of millions of years. Historical crop failures can be linked to genetic limitations (see Figure 6).

The potato blight that caused mass destruction to the potato fields in Ireland during the middle of the nineteenth century can be attributed to diversity loss. For approximately 250 years, the potatoes in Europe descended from just two crop varieties. Potato blight, caused by a fungus known as *Phytophtora infestans*, spread throughout Europe by following the path of genetic susceptibility in these two varieties. But such losses were preventable had sufficient diversity been maintained.

The potato originated in cultivars from the Andes in South America, where hundreds of different varieties of the potato are grown. The blight that struck Europe and Ireland also took hold in the distant Andes. In Ireland 2 million people died from starvation, whereas in the Andes, only a few crops were lost. There, potato varieties survived due to the presence of genes that conferred resistance to the blight. After the epidemic, Andean wild potato relatives were widely used to restock European potato farms.

Other epidemics among genetically uniform food crops include brown spot disease in an Indian rice crop that began the infamous Bengal famine, a wheat epidemic in 1917 in the United States, and an oat crop failure in the 1940s which eliminated eighty percent of the crop. Each time an outbreak occurred, resistant forms were

Figure 6. Crop Diseases Resulting from Monoculture

YEAR	DISEASE	CROP	COUNTRY/ REGION	AMOUNT OF CROP DAMAGE	$ VALUE
900	Viral	Corn	Central America	...	...
1845	Fungal	Potatos	Ireland	1 million died of starvation	...
1860	Fungal	Grapes	Europe	...	...
1865	Fungal	Coffee	Ceylon	...	...
1890	Viral	Sugar Cane	Indonesia	...	...
1916	"red rust cut"	Wheat	U.S.	...	...
1954	"red rust cut"	Wheat	U.S.	75%	...
1969	Bacteria	Rice	Asia	...	...
1970	Virus	Rice	Phillipines	...	...
1970	Southern corn-leaf blight	Corn	U.S.	15%	1 Billion
1984	Citrus Canker	Citrus	U.S.	18 million trees destroyed	...
1989	aphid/ insect	Wheat	U.S.	34 million acres	$300 million

Graphics by Platt & Company

obtained from the centers of diversity containing wild relatives of the crops that failed. These resistant forms were essential to reestablishing food production and insuring the survival of the species.

Crop Uniformity

The increasing trend toward uniform farming practices that encourage growing one variety of a crop on a mass scale is almost certainly going to be exacerbated by the availability of transgenic seed. The resulting crop patterns may well increase the fragility of the crop and permit the introduction of widespread disease. This is likely because genetic diversity is integral to sustainability, balance, and the survival of crops from harvest to harvest. The genes for adaptations favoring survival evolved under conditions of diversity. Wild plants are under constant pressure from pathogens, pests, severe climates and unfavorable soils. As a result, they have evolved a myriad of strategies for survival including thorns, natural toxicity and fibrous tubers.[6] Many of these defensive characteristics, maintained as part of the reservoir of genetic diversity, are being progressively lost through domestication.

Genetic engineers drive this process still further by isolating a small subset of these traits and putting them into a selected small number of cultivars. For instance, the Liberty Link™ technology is currently in a limited number of seed stocks, greatly reducing the seed types available for sale just one or two growing seasons ago. This trend amplifies the selection pressures towards uniform traits. According to some market surveys, many consumers "want" only limited varieties for their home pantries. Such pressures further compromise the genetic variation normally maintained by natural selection. Today, protecting genetic seed banks and their associated diversity is a diminishing enterprise, maintained by only a few committed scientists who run germ line seed banks on shoestring budgets. It is worth reviewing the history of germ plasm conservation to understand the magnitude of what may be lost if we permit the genetic homogenization intrinsic to mass production of transgenic crops.

Introduction of Pesticides

Crop uniformity carries hidden costs in heightened needs for protection. As the domination of smaller and smaller numbers of corporation selected, transgenic varieties increase, so does the risk of catastrophic disease or pestilence similar to those cited above. This dilemma, in part, encourages the proliferation of the chemical means of disease and pest control.

The pattern of increasing pesticide reliance to control insect infestations is disturbing. And chemicals do not appear to be the answer. In the last forty years, the percentage of the annual crops lost to insects and disease in the United States has doubled. Since pesticides were introduced in the 1940s, the proportion of crops lost to insects in the U.S. has grown by 13 percent.[7] By 1945, farmers were using 200 million pounds of pesticides each year in the United States.[8] Thirty years later, the annual total had risen to 1600 million pounds. Currently, global pesticide sales continue to rise. In 1996 global sales of pesticides topped $30.5 billion and are predicted to rise to $33.1 billion by the year 2001.[9]

Farmers are now not only committed to buy new seed each year, but also a set poundage of these pesticides, especially those genetically predicated on their seed purchases. With the advent of Liberty Link™ and Roundup Ready™ technologies, this pre-commitment will be even stronger. Transgenic scientists appear to be ignoring the maxim that just as insects develop immunity to pesticides, disease causing organisms also adapt to chemicals and the genetic defenses of plants. All of these organisms have coevolved by adapting to a changing environment. By losing plant varieties, plant growers are losing opportunities to breed for essential natural defenses.

Risks of Monoculture and Monopoly

We have other concerns about this disturbing trend towards increased control over genetic diversity. When a genetically controlled monoculture of a given crop is substituted for a race of microacclimatized potatoes, corn, cotton or soybeans, this substitution imposes on the farmer a higher dependency on uniform soil conditions, higher fertilizer and water use, and machine-dependent harvesting methodologies. More critically, any blight, fungus, rot or other disease which might previously threaten a portion of a region's crops now may threaten to devastate a swath of genetically identical cultivars. The likelihood of some such apocalyptic scenario might be expected to motivate corporate concern to anticipate and monitor transgenic crop loss. Instead, an entire army of detail men have been assigned only to monitor seed use and prevent theft of transgenic varieties. This omission left an entire industry unprepared to cope with the first two major transgenic crop failures.

Genetically Engineered Crops in Jeopardy

The first occurred when the cotton bollworm successfully overran Monsanto's Bollgard® crops in Texas during the summer of 1996, an event reviewed above in Chapter 4. In 1997, a disturbingly similar failure afflicted a portion of the Roundup Ready™ cotton crops in the Southeast.

In the early weeks of August, 1997, farmers throughout the mid-south region of the United States watched the cotton bolls on their Roundup Ready™ cotton fall off. The resulting damage ran into the millions of dollars affecting at least 60 different farms. The failing plants, containing an inserted Roundup Ready™ gene making the cotton plants able to withstand two seasonal dousings of Roundup® herbicide, were among the first to be grown commercially. The 1997 planting season was to be the debut of this much heralded product. Approximately 600,000–800,000 acres of the newly bioengineered crop created by Monsanto Company were sown with Roundup Ready™ cotton, or about 2.5% of the 14 million acres of cotton planted nationwide.

But three-quarters of the way through the growing season, something went awry. Cotton bolls, the billowy fruit of the plant which embraces the cotton seeds (which are ginned from the raw fibers) were lost after the second and final Roundup® application. Many of the bolls simply fell off of the plant after spraying. These apparent failings occurred in the states of Mississippi, Arkansas, Tennessee, and Louisiana, and were thought to be Roundup® related. According to Robert McCarty of the Bureau of Plant Industry in Mississippi, whom we reached at the peak of the epidemic on September 2, 1997, "we are receiving complaints from farmers every day." The complaints were all identical: the bolls became deformed and subsequently fell off the plant. Mr. Bill Robertson, a cotton specialist in Arkansas, pointed out that they experienced similar problems as well. "We call the malformation 'parrot-beaked', because the bolls look like the beaks of parrots, then they fall off of the plant before they are mature," Robertson said. The first reports of the apparent failure of the crops placed the number of affected acres in the 4–5000 range, though according to Mr. McCarty, "we are talking at least 20,000 acres in Mississippi alone, and we are getting new complaints every day. Now that is a lot of acreage, economically speaking. Some farmers are individually losing $1 million due to this problem."

As of late 1997, the investigation of this disturbing reversal of fortune remains inconclusive. At the state level, agriculture agencies are gathering data mostly for economic analysis. These agencies support the farmers, and they became involved in order for the farmers to gain compensation for their losses. Monsanto Company is also doing an investigation of economic losses, and is likely to be the only one capable of discovering why the Roundup Ready™ cotton crops apparently failed. The likelihood that they will finger their own genetic contribution to the loss is low. According to Karen Marshall in Pubic Affairs of Monsanto Company, "there are a number of environmental factors that can put stress on cotton plants."

Beginning in August 1997, the apparent failures did not occur in all cotton varieties in the same region, just in those few varieties that were Roundup Ready™.[10] Ms. Sunny Jeter, a Roundup Ready™ marketing representative of Monsanto, insisted that the apparent failure was only occurring in a very small portion of the Roundup Ready™ cotton crops. She emphasized Monsanto was being very proactive in getting information to farmers about the problem. Tommy McDaniel, a State of Mississippi agricultural specialist acting on the front lines, took a different tack when we interviewed him. McDaniel declared, "Monsanto is not talking to anyone and they are not saying what is causing the problem."[11]

The details that have emerged to date give little cause for optimism. Something appears to have gone wrong with the Roundup Ready™ technology. The apparent failure is occurring in specific Roundup Ready™ Paymaster (a Monsanto subsidiary) varieties #1244, #1215, #1330, and #1220. All of these varieties were used in the two previous years without any apparent problems. But in 1997, the Roundup Ready™ versions of these varieties apparently failed.[12] Several extension agents and investigators with whom we spoke speculated about why the crops are apparently failing. Most are hypothesizing that the newly inserted gene has caused instability within the Roundup Ready™ cotton crop genome, an effect made evident in the F_3 to F_5 generation of the plant.

As with the apparent failure of the genetically engineered *Bt* cotton, the 1997 Roundup Ready™ apparent failure may remain unsolved. Because of the lack of tracking and effective plant epidemiology, we may never know its causes or origins. Monsanto's view, in the fall of 1997, was simply that "the information is not available." The government does not require reporting after deregulating the crop, leaving the public and the farming community alike in the dark about the true cause of the problem.

We see a larger issue here. When Monsanto released its technology in 1997, they asserted it was "ready" for commercial scale application. But in the first year of large scale planting of two of its major crops, a significant portion of the released crops apparently failed. Should not geneticists have studied just where in the plant's genome its new gene was inserted? What occurred in the plant to make it shed its seed bolls prematurely? Should not a tracking system and "hotline" have been in place?

Comment

These disturbing events, although limited in scope, underscore our concern that mass plantings of transgenic crops are at the least premature. If these engineered plants were any other life form, no one would have permitted their widespread introduction into the environment without an Environmental Impact Statement. Scientists still do not have the answers to fundamental questions. They do not know why certain genes "take" in their new host and others do not, or where the gene goes once it is ensconced in its new home. These questions are even more difficult to understand with crops such as cotton. Most crops manufactured today are hybrids. In most cases the first generation of seed is a known genetic entity when it is planted. Cotton is not a hybrid. The seeds that are planted often represent the fifth or sixth generation of plant descendants. With each new generation there is a reorganization of the genes. A new gene may be effective in one place within the genome but may cause another quite different reaction when reorganized the next year. In other words, Monsanto, the seed companies, and the farmers do not know for any given year where the new gene has become integrated in that year's genome, or how exactly it will affect the plant growth. Planting non-hybrid, genetically engineered plants one year after another can be very much a form of roulette.

We are thus left with disturbing questions as transgenic crops go into mass production. How much are we willing to jeopardize the evolutionary future of our food crops? How much uncertainty is generated by transgenic creation of new plants? And are we really ready to let large corporations play God in the critical area of food biotechnology?

---

[1] Ellstrand, N. C. 1988. "Pollen as a vehicle for the escape of engineered genes?" In: J. Hodgson & A. M. Sugden, (eds.) *Planned Release of Genetically Engineered Organisms* (Trends in Biotechnology/Trends in Ecology & Evolution, Special Publication) Elsevier Publications, Cambridge. pp. S30–S32.

[2] Charles Connor, "Roundup™-tolerant corn seed ditched," *Memphis Commercial Appeal* 14 November 1997; Section 1.

[3] Reuters, "Pioneer will not carry Roundup Ready™ corn," Des Moines, Iowa, 13 November 1997.

[4] Reuters, "Monsanto seeks widespread Roundup Ready™ corn use," St. Louis, Missouri, 13 November 1997.

[5] Economic Research Service, USDA. *Forces shaping US Agriculture.* July 1997.

[6] Paul Raeburn, *The Last Harvest; The Genetic Gamble That Threatens to Destroy American Agriculture,* University of Nebraska Press (Lincoln),1996, p. 96.

[7] Fowler and Mooney, p. 48. The explanation why pesticide use is on the rise has much to do with the role of the pesticide. Many pesticides alleviate all pests, beneficial and harmful. Beneficial pests feed on the pests that are harmful to crops. Laws of nature have created settings whereby there are more "harmful" pests than "beneficial," otherwise beneficial insects would die off from lack of food. Pesticides generally do their job by killing everything, though their task is often short-lived. Typically, when the pesticide dissipates, harmful insects return to their food source while multiplying rapidly for their survival. Many surviving pests include those resistant to the chemical that killed their ancestors, redoubling our work. Dr. Carl Huffaker asserts that "when we kill a pest's natural enemies, we inherit their work."

[8] Agrow: World Crop Protection News. 13 December, 1996.

[9] Agrow: World Crop Protection News. 13 December, 1996.

[10] Davis, Keith, Bureau of Plant Industry, Mississippi. Telephone interview. 16 October, 1997.

[11] McDaniel, T. Telephone interview. 4 September, 1997.

[12] McCarty, Will, Cotton Extension Specialist, Starkville, Mississippi. Telephone interview. 6 October, 1997.

9

Ethical Issues and Long-Term Consequences

The occurrence of crop failures, lawsuits, and disgruntled consumers underscores a more fundamental issue: no one has fully explored the ethics of moving this technology so rapidly into mainstream commerce. Many scientists, government officials, and corporate lawyers have assumed that because "only plants" were involved and no immediately obvious issues of safety were at stake, no ethical issues were involved with transgenic food crops. Nothing could be further from the truth.

The ethical issues generated by agricultural biotechnology are both simpler and more complex than those in animal or human genetic engineering. They are simpler because alterations in plants do not directly affect people or their well being. No one is talking about making Frankensteins from transgenic crops. Unlike human genetic engineering, transforming a plant does not alter its soul or affect the basic elements of its humanity or culture. From a genetic perspective, most alterations created by inserting single genes into transgenic plants are hardly radical in terms of the proportion of the genome affected. Indeed it is difficult to compare the radical transformation of plants to similarly extreme alterations in animals. Animal rights advocates are understandably outraged by the creation of growth hormone transformed hogs into arthritic, clumsy and distorted versions of their "natural" hog-ness. Few members of the public have been comparably disturbed by the creation of a slow ripening tomato.

The traditional breakdown of ethical issues in genetic engineering questions the rightness or wrongness of a particular intervention. Most analyses to date come from the perspective of *positive utilitarianism*. From this philosophical viewpoint, genetic engineering is assessed in terms of its benefits arrayed against its risks and costs. At a superficial level, this analysis almost always leads to a high valuation of agrobiotechnology, because the "good" consequences (i.e., higher crop yields, lower reliance on pesticides, and improved economic returns) appear to far outstrip the largely remote and distant risks from monocultures and disease susceptibility. And even though the start-up costs of biotechnological inventions are steep, the returns in the agricultural sector have quickly outstripped those investments.

But this analysis leaves much to be desired by way of ethical perspective. It is equally valuable to consider the consequences of *negative utilitarianism*. Here, the major concern is to offset or mitigate present or future harms, and to ask if the technology affords new ways of doing so. For instance, the good of agrobiotech could be assessed against the likelihood of averting starvation, or assuring a livelihood for otherwise impoverished farmers in the developing countries.

A third way of examining biotechnological advances in agriculture is to ask whether or not they raise *teleological* questions about the basic rightness of genetic alteration of living things. Are there any reasons, a teleologist might ask, to question whether transgenic plants are natural or their creation a subversion of some overarching natural order? A fourth perspective is that of *deontological* ethicists who are concerned about rules and limits on human regulation. Persons holding this baseline approach may question the growing trend to permit modifications of plants without special regulations. Proceeding without rules or controls only makes sense if transgenic plants are in fact no different from the natural forbearers, a biological and philosophical distinction we dispute.

While we will explore each of these viewpoints below, they miss the major issue in agrobiotechnology: the consequences of biotechnological interventions on a mass scale. Under such scalar interventions, decisions are being made that may irrevocably change the composition of the food supply for millions of people. The fact that most of the affected populations have had little or no participation in the decisions affecting their diets—and ultimately, their lives and livelihoods—is an ethical issue in its own right.

As we saw in the Introduction, in Europe this reality is particularly troubling. Engineering plants is considered by many Europeans to resemble the same manipulations that led to engineering animals. Even here, some people believe plant genetic engineering is playing God, desacralizing nature, and creating new organisms through genetic tinkering. While such concerns are outside the realm of science, they are worthy of close examination and possibly respect.

Sanctity of Life Arguments

While some bioethicists have promoted the idea that the human genome carries intrinsic worth and that people have "rights" to the integrity of their genetic heritage, such positions have proven difficult to ground in traditional ethical theory or jurisprudence. How much more difficult then to assert a similar claim when arguing we should not be manipulating the genome of plants. Many agricultural practices have already changed the form and genetics of plants through selection for millennia. It is therefore largely ineffective simply to argue we are changing the form of plant life in a fundamentally novel way through transgenics, as grounds for opposing food crop genetic engineering.

But in the *choices* of transgenes, human selection is clearly extending and perhaps overriding natural selection. Genetic engineers draw from a much larger gene pool (including the genomes of exotic species) than does nature. Plant scientists practicing transgenesis can in theory have a broader and more lasting impact on any targeted plant than will natural selection. The ethical issue is not merely that transgenic crops are "unnatural" to the extent they are built from exotic genes, but that human selection has created a new future for the plants' progeny that reinforces certain private and economic values over natural, aesthetic, or human nutritional ones. Privatization is thus at the core of ethical issues.

For the first time, transgenic science has enabled single corporations to claim an entire novel genotype as their own. The privatization of gene stocks violates and limits the natural order by making genes private property. While hybrid gene-derived plants (e.g. roses) have been granted patents before, their *genes* were not the exclusive province of their maker. Anyone could (and did) use a commercial variety to outcross with their own to make a novel plant once again. It was only the cutting and propagation of a clone which was outlawed by the patent provision. Transgenic seed, even when admixed with new genetic material by hybridization or outcrossing, *remains* the property of its maker, as long as the transgene is retained.

Genetic engineering also raises major scientific issues of possible genome destabilization and ecological disruption (to be discussed below). It is also possible to argue the creation of transgenic crops is intrinsically immoral because it violates the evolutionary integrity of the organism. Questioning the exotic origin of a gene in transgenic plant species is valid because plants have rarely, if ever, acquired genes across genera lines in the past. The profaning of the natural order by transgenesis is a widespread and pervasive sentiment in some religious circles.[1]

Scale of the Enterprise

Natural gene transfer can occur across and within species and perhaps even between phyla (note the bacterial origin of chloroplasts or mitochondria). Such realities reduce the moral cogency of the fact that humans have "unnaturally" moved other genes between plant and bacterial species. What we find relevant is the scope and scale of this engineering. At a certain scale, genetic changes in one plant species can transform nature by altering the balance between species, and thereby disrupting ecological stability. Even in agriculture, encouraging a monoculture can confer man-made dominance to one cultivar over another and (as we saw) increase the vulnerability of a food crop to pestilence. Resulting losses of genetic diversity for resistance genes can put future generations at risk. Assuming the majority of transgenic crops will be put into agricultural settings in which large-scale disturbance and loss of control already occurs sporadically, the major ethical issues revolve around any features exclusive to transgenic crops.

The scale of an undertaking in which whole plantings are being made with one genetic line eclipses any past use of genetics by several orders of magnitude. With transgenic crops, literally millions of individual plants are being created with identical genetic alterations. A mistake made in one, say vulnerability to blight or disfigurement, will be repeated in all of the rest. While large plantings of hybrid varieties have been made in the past, the scale of crop operations using Roundup Ready™ soybeans or corn is so huge—estimated to go to almost 100% of all crops planted in the year 2000—that these crops will inevitably impact the world food crisis (a point discussed in the previous chapter) and shift patterns of agriculture significantly. Although several different genetic backgrounds will be used, all plants will contain an identical gene. It is also likely that the initially attractive package of yield plus the simplicity of pesticide usage touted in crop management brochures of transgenic crops will prove seductive to many farmers.

Incentives and Freedom of Choice

Ironically, as transgenic crops become marketable they offer farmers short-term, economically attractive choices while restricting those of consumers. For growers of soybeans or corn, the availability of Roundup Ready™ technology provides an opportunity to select crops for mass production and spraying largely outside of governmental reach. In part, this latitude is created by current legislation and a great relaxation in regulatory oversight. Previously, farmers set aside acreage for tilling certain crops or allowing other acreage to lie fallow. This rule has been replaced with a more laissez-faire regulatory climate. Since The Freedom to Farm Act of 1995, the government has permitted farmers a much wider range of choice than they had under the prior Acreage Reduction Program. Under the latter program, the government offered farmers incentives to retire productive land from crop production. Under the Freedom to Farm Act, land previously taken out from production is now released for planting. As one of our farmer sources told us, the biggest boost to agriculture in the last two years was the increase in freedom to choose.[2] This enhanced autonomy translates directly into new dollars proportional to the economic value of the crops that could be grown. In the instance of our farmer source, for economic reasons, he elected first to grow transgenic cotton and secondarily, Roundup Ready™ soybeans on all his previously reserved crop lands.[3] With the subsequent crop failures, we suspect he is more dubious of his choices.

The previously circumscribed freedom of choice is now left to the individual farmer. The decision to use a transgenic crop will be driven almost entirely by market forces. The net effect of these forces will increase the land planted with transgenic crops affording high rates of return for investment.

Transgenic seed producers stand to benefit substantially from this shift. The Freedom to Farm Act frees up as many as 4.8 million acres previously taken out of production. Given the initial economic advantages of Roundup Ready™ and related engineered crops that require less expensive chemical use, the net effect of this act is to create an unprecedented economic opportunity for transgenic crops. This reality is not lost on corporate entrepreneurs who are promoting genetic engineering.

Based on what was taken out of production, academics project the majority of this freed-up acreage will go to wheat and soybeans. With each increment in increased acreage, substantially more soybeans and wheat will be planted. Although wheat acreage may only increase marginally with a low return, with the maximal increase in freed up land, wheat acreage could increase up to 6.5%. Likewise, soybean acreage will grow by an estimated 6.9%, representing some 4.1 million additional acres. The additional production will largely be exported—increasing the pressure on recipient countries in places like the European Union to permit genetically engineered crops to enter their commodity trade.

Given the availability of transgenic crops, it is likely that farmers will grow only those crops that are most economically viable. The domination of soybeans in the last decade will be intensified, and the resulting crop pattern will likely overproduce this commodity. In our analysis, this trend is likely to have a counterintuitive effect: greater availability of transgenic soy protein may actually decrease human soy consumption. The full utilization of soy protein in human foods is likely to be limited by two factors: the increased demand for animal feed supplements and pharmacological questions raised about soy in the human diet.

It is likely that any economic benefit of transgenic soybean products will lead to greater use of soy commodities in animals. With increased efficiency, the use of soybeans primarily (i.e., at 90-95%+) for animal fodder will increase or stabilize, providing a cheaper foodstuff for animal production.[4] The ability of livestock to reduce the food value of soybeans by about 8.5 to 1 means that a significant portion (approximately 85-90%) of otherwise edible food will be lost for human consumption. Additionally, as we have shown, there is no assurance that yields from Roundup Ready™ crops will be any greater than the yields of its non-transformed predecessor. So, contrary to representations by industry leaders like Monsanto assuring consumers that transgenic crops guarantee world human food stocks will go up, it is equally likely that human available protein production will go down. Even if a significant portion of the soy crop were made available for human consumption, questions have been raised about the suitability of soy based protein as an exclusive dietary protein source. More to the point, ethical issues about consumer choice remain.

Informed Consent

Biotechnology clearly introduces novel issues of informed consent because it introduces new substances into whole foods, transforming ordinary fruits, vegetables, grains and meats into something they were not.[5]

While the crops produced are new, there is little scientific evidence proving they are intrinsically unsafe. According to Texas A & M professor Paul Thompson, "there is an unusually high degree of consensus among toxicologists, nutritionists, and applied biologists that properly conducted gene transfers pose considerably less risk to consumers than [do] conventional chemical and breeding techniques for developing or modifying foods."[6] From our vantage point, this statement seriously underestimates the secondary health risks of transgenic crops and ignores the three prior issues we have identified: 1) transgenic crops do pose ecological and agricultural risks; and 2) mass application of biotechnological innovation will shift reliance towards certain selected chemicals that themselves pose novel risks; 3) biotechnology promises to dislocate major cultural values and creates a de facto acceptance of human dominion over nature. Primarily these risks grow out of the scale of the enterprise.

Ecological Risks

The conscious choice of a few genes for mobilization and widespread replication substitutes human judgment for natural selection. From a theological viewpoint, it is questionable that the agribusiness scientific staff have the collective wisdom to determine what constitutes the "good" when it comes to desirable genes. The fact that their choice could become self-sustaining (e.g. if the gene escaped into the wild) is cause for further concern. Initially, this and other adverse impacts potentially resulting from mass scale transgenic operations are likely to be invisible. One potentially insidious effect of reliance on genetically engineered herbicide resistant technology is the repeated use of single herbicide preparations. The repeated applications of a controlled sequence of four or more different herbicides typical of pre-transgenic farming could be expected to transiently affect soil micro-organisms. But the sustained reliance on a single herbicide such as glyphosate or bromoxynil would predictably shift the soil microflora for longer periods, perhaps changing the overall composition of the soil's living matter irrevocably. Such an effect, should it occur, could affect soil quality for future plantings, particularly since germination in some herbicide treated soils has been reported to be impaired. Here the ethical concern is responsibility for future generations, a point taken up below.

Other second order effects are more noticeable and noteworthy. Among these are the reduction in the type and availability of non-engineered foodstuffs and the resulting loss of consumer choice and preference. The concept of voluntary choice as a driving force is termed "consumer sovereignty." Consumer sovereignty is derived from the ethical principle of autonomy which declares each person has an intrinsic right to participate fully in decisions that affect his life choices. As we have shown, the loss of sovereignty over the selection of non-transgenic produce has already occurred. At the core of the debate is a variation on the ethical maxim of informed consent: the legitimacy of consumer assertions that they have the right to exercise an informed choice.

Informed consent requires that individuals be given opportunity to make significant, life-affecting choices. Ethical standards established in the medical context over the last 50 years protect individuals from coercion, fraud, deceit, or undue suasion in a variety of circumstances. Patients must be afforded procedural safeguards to assure that choice is freely given and informed.

Those who argue transgenic foods are unworthy of being given comparable standing to patients are ignoring widely held religious, aesthetic, moral and cultural beliefs. As we have shown, consumers have good reason to believe that transgenic food crops may be qualitatively different from non-transgenic ones. Such foodstuffs often contain novel gene products and are produced by external manipulations in concert with natural means. Their mode of creation circumvents traditional agricultural norms, especially those of organic food production. For many, especially the very religious, the objections to this mode of production is both aesthetic and principle based, and not merely a distaste for new technologies.

Health Risks?

While we have emphasized the indirect health risks from herbicide residues posed by transgenic foods, other bona fide safety issues have been virtually ignored. Pro-transgenic spokespersons, including USDA's Daniel Glickman, as well as Monsanto's representatives, have insisted that Roundup Ready™ soybeans and related products are safe. This view is clearly shortsighted. By focusing solely on chemical residue data, all of which are within EPA tolerance limits, these persons have missed a key fact: to create Roundup Ready™ soybeans, Monsanto's genetic engineers boost the activity of a gene which makes critical amino acids like tryptophan. The fact that these amino acids are in greater abundance inside plant cells not only confers resistance to Roundup Ready™, but arguably shifts plant metabolism. One of the main byproducts for soybean metabolism is a class of compounds called isoflavonoids. These plant substances, because they have a remarkable similarity to our body's own estrogens, are called phytoestrogens. Their increased presence in transgenic soybeans could explain the observation of higher fat content in milk from transgenic soy-fed dairy cows.[7]

Estrogenic substances play critical roles in controlling sexual differentiation, calcium metabolism, immune function, carcinogenesis, and blood clotting.[8] Hence, it is essential to know how much or little phytoestrogen is present in soy products. The recent non-industry revelation that Roundup® treated legumes may have elevated phytoestrogens,[9] if confirmed, provides disturbing evidence that transgenic crops are not only different, but may well have a dramatic impact on the health and well-being of those who rely on soy protein as a major part of their diet.[10]

Nondairy infant formulas are a case in point. New data on hormonal activity of soy components suggest the medical community may have to limit or at least reevaluate reliance on soy-based products early in life. The culprit is the remarkably high estrogenic activity of soy isoflavones. If ingested by nursing infants, these isoflavones can produce circulating levels equivalent to 13,000 to 22,000 times the normal plasma estradiol concentrations found in babies, with unknown and potentially dangerous secondary effects.[11] (Early exposure to estrogens, notably diethylstilbestrol, is associated with sex organ dysfunction and an increased incidence of certain endocrine responsive tumors such as vaginal adenocarcinoma.) While other effects of isoflavone-rich soybeans may be healthful, notably in reducing breast cancer risk and reducing cholesterol levels, a diet high in soy may have detrimental effects. Among these effects are the possibility of adversely affecting reproduction and disrupting the menstrual cycle.[12] Given these prevailing concerns, creation of only weakly estrogenic soy products would appear to be a desirable goal for genetic engineers. The fact that no overt evidence of such a trend exists among major transgenic producers is another cause for concern.

Consumer Sovereignty

The special circumstances of transgenic food crops provide a basis for reasserting consumer rights to sovereignty in their food choices. These reasons include emotional reactions against transformed or otherwise "unnatural" foodstuffs; religious objections to the transspecies movement of genes; and wishes to support the "right livelihood" of those food producers, such as organic farmers, who choose non-technology based methods of production.

Presently, consumers lack any tool which would allow them to discriminate between transgenic and non-transgenic foods. The resulting denial of free choice is partially the result of the Food and Drug Administration's reluctance to require regulatory review of genetically modified foods except where a novel substance is introduced into the human food chain. Since for the most part, the genes introduced into plants could be found in other plants or their bacterial hosts, the argument has been that *totally* new substances are rarely if ever found in transgenic crops. The lack of regulatory activity combined by the introduction of transgenic crops under ambiguous labels (e.g., the "McGregor Tomato") further erodes the exercise of consumer preferences.

Honest Marketing?

The McGregor Flavr-Savr® tomato by Calgene, Inc. was the first engineered food crop to be introduced into the commercial market. The public was made aware that the Flavr-Savr® gene technology was to be used in commercial tomatoes. The media carried stories about their enhanced ability to stay fresh on the vine due to the introduction of a gene that delayed its ripening capacity. But U.S. citizens remained skeptical about the value of these expensive, engineered tomatoes, and they floundered in the marketplace. Calgene was forthcoming about the failure, acknowledging that the tomato's altered genetics may have led to public disapproval.

But its subsequent actions were troubling. In response to the consumer wariness of the Flavr-Savr® tomato, the new owner of Calgene, Inc., Monsanto, re-marketed the tomatoes under the name "McGregor" without reference to their genetically engineered origins. When they first began buying interest in Calgene, Inc., Monsanto gave Calgene a company that they owned, Garguilo LP, the largest U.S. tomato grower.[13] When Monsanto completed the buy out of Calgene, Inc. in January 1997, they reacquired Garguilo and began production of the "McGregor" and the "Garguilo Farms" tomatoes concurrently.[14] These brand names of course obscure the reality that the McGregor tomato has an altered genome. The apparent lesson learned from the introduction of Flavr-Savr®, as seen through industry eyes, is the value of nondisclosure. If disclosure of genetically engineered status causes "irreparable harm" for Flavr-Savr®, the marketing answer lies in nondisclosure the next time around.

The present wave of engineered crops is entering into our food supply in a much more insidious way than the much heralded Flavr-Savr®. Industry has decided to silently invade food market shelves by denying any visible identifiers of genetic engineering. If there is a label on transgenic crops, it is only on the seed bag seen by the farmer under euphemistic names such as Bollgard®, Endless Summer® and New Leaf®. The net effect is to subvert the normal process of consumer choice by suppressing the knowledge needed to freely choose. The cornerstone of such a privilege is labeling.

---

[1] Hoban, T. J. and Kendall, P. *Consumer Attitudes about Food Biotechnology*, NC Cooperative Extension Service (Raleigh), 1993.

[2] M. R. Dicks, D. E. Ray, D. G. de la Torret Ugarte and R. L. White, "The Freedom to Farm Act," The 1995 Farm Bill, No. 6 pp. 1–8, October 1995.

[3] Zanone, George, farmer. Personal interview. 19 June, 1997.

[4] Frank Flighter, edible oils consultant to the United Soybean Board. Telephone interview. 26 February, 1997.

[5] Paul B. Thompson," Food biotechnology's challenge to cultural integrity and individual consent," *Hastings Center Report* 4: 34–38, 1997.

[6] Ibid.

[7] Hammond, B. G., *et al.* "The feeding value of soybeans fed to rats, chickens, catfish and dairy cattle is not altered by genetic incorporation of glyphosate tolerance," *Journal of Nutrition* 126 1996: 717–727.

[8] Ingram, D., Sanders, K., Kolybaba M., and Lopez, D. "Case control study of phytoestrogens and breast cancer." *Lancet* 350: 990–997, 1997.

[9] Sandermann, H., E. Wellman, Bondesministerium für Forschung und Technologie, *Biologische Sicherheit* (1988): 285–292.

[10] We are currently conducting a controlled test of the phytoestrogen-glyphosate link. Preliminary data suggest that genetically engineered soybeans are, if anything, slightly lower in overall estrogen content.

[11] K. R. R. Setechell, L. Zimmer-Necehmias, J. Cai and J. E. Heubi, "Exposure of infants to phyto-oestrogenis from soy-based formula," *Lancet* 350: 23–27, 1997.

[12] A. Cassidy, S. Bingham, and K. D. R. Setchell, "Biological effects of soy protein rich in isoflavones on the menstrual cycle of premenopausal women," *American Journal of Clinical Nutrition* 93: 225–[]233, 1987.

[13] J. Riggins, Calgene Incorporated, Oils Division. Telephone interview. 11 June, 1997.

[14] A. Thayer, "Monsanto gets all of Calgene," *Chemical and Engineering News*, April 7, 1997.

10

Labeling

If the remedy to consumers "right to know" is labeling, few if any food manufacturers are availing themselves of the cure. The argument they give is that "they can't." Why? Once the transgenic food product reaches the open market, virtually all of it has been mixed with non-engineered conventional product. According to Pillsbury foods, genetically engineered soybeans are found in virtually all of its food products because transgenic soybeans and natural soybeans are already commingled when they buy them as commodity.[1] As we have shown, the routine mixing together of transgenic versus non-engineered food staples virtually guarantees that the end product cannot be accurately labeled. The combination of engineered and non-engineered product has taken place in spite of a recent industry poll which found that 93% of all American citizen respondents deemed labeling highly desirable or necessary.[2]

Even though the public appears to want labeling, they may not have that wish granted. A recent example where citizens were denied this right occurred when the people of the state of Vermont voted for labeling of dairy products that contain a growth hormone given to dairy cows. The growth hormone is called recombinant bovine growth hormone (rBGH), and is manufactured by Monsanto Company under the trade name Posilac®.

Regardless of the democratic vote substantiating the necessity of labeling products containing additional hormones, on August 8, 1996, the U.S. Circuit Court of Appeals ordered Judge Garvan

Murtha to issue an injunction to block Vermont's rBGH labeling law. This decision came after the Dairy Association, with Monsanto in the lawsuit as a "friend of the court," filed a lawsuit asserting the existence of a First Amendment right "not to speak."[3] This order continues to conceal the fact that a rapidly growing amount of our dairy products contain rBGH, thwarting the will of 93% of the American public who would like labeling of these products.[4]

This attack upon consumer sovereignty is being fueled by the industry's reluctance to permit labeling of genetically engineered foods. The industry position is that any such label would stigmatize otherwise wholesome products and would provoke unwarranted consumer anxiety and fear about food safety.[5] Secondary concerns such as the cost of labeling and its proportionality to the public health risks posed by mislabeling have also been raised.

In our view, this position is both dangerous and shortsighted. If transgenic products are as much or more wholesome than are ordinary foods, then let industry make its case openly to the public. If public fears are exaggerated, let industry dampen them with the kind of educational and advertising campaigns they have perfected over the years to sell us Nutrasweet®, Saccharin®, Simplesse® and other synthetic food products. Without labeling, of course, all of these admittedly expensive efforts would be unnecessary. But the recent ruling to require disclosure of generally useful, health related information of food content even when it provided data which only a few consumers would find relevant (e.g., diabetics or phenylketonurics) sets an important precedent.

What's Happening in Europe

The European Commission has proposed that companies must label products which "may contain or may consist of genetically modified organisms."[6] The European community has been very concerned about the absence of labeling on products that are genetically engineered, specifically soybeans and maize. In January 1997 the Commission agreed upon definitions for novel food regulations that took effect in May 1997.[7] In the definitions, genetically engineered food products must be labeled and sold separately.[8]

Taking a further step, Austria and Luxembourg have banned the import of genetically engineered maize. The governments of these countries believe that the maize crop engineered by Ciba to contain the toxoid *Bacillus thuringiensis* also contains a genetic marker conferring antibiotic resistance which could be transferred to humans or animals. In 1997, 20% of the Austrian population, or more than 1.2 million people, turned out to sign a petition calling for a ban on genetically modified foods and on patents on genetically modified plants and animals.[9] This was the second highest turnout in their history of national petitions and the highest ever for a grassroots initiative. In this same year, Denmark decided that all foods containing genetically engineered soybeans must be labeled. In Germany, the Green Party has been working diligently to block engineered food products. Switzerland has passed laws stating that all food products containing genetically engineered soybeans must be labeled. In a corporate decision made in April 1997, the chocolate maker Kraft-Suchard recalled 500 tons of chocolate after genetically engineered soybeans were detected in the soy lecithin which acts to bind fat and sugar.[10]

Reasons

The opposition in Europe to anything "genetic" or "contaminated" may be due to their recent experience with "mad cow disease." In 1996–97, 1.5 million cattle were incinerated after they were suspected of carrying bovine spongiform encephalopathy (BSE) that killed 18 people in Europe. Within the last 10 years, the population of Europe has also learned more of the academic support for German World War II atrocities, where scientists, doctors, and other professionals were implicated in using genetic ideology to justify heinous human experimentation without proper informed consent. The European general public, therefore, may be more sensitive about the need for disclosure. Perhaps in part because of these historical precedents, Europeans have adamantly demanded their right to choose or reject genetically engineered crops. A case in point highlights the intensity and scope of this debate.

Novartis, the new Swiss agricultural biotechnology firm formed in 1997 by a merger between Ciba-Geigy and Sandoz has created an engineered corn crop known as Maximizer® corn. Maximizer® contains the *Bt* gene and is touted by Novartis as a crop which will produce maximal yields. But in countries like France and Austria, people are demanding labeling or even a ban on the importation of Maximizer® corn. The European Parliament (EP), an advisory panel for the European Commission, met on April 8, 1997, to discuss the dilemma of importation of genetically modified maize. The EP voted to disapprove transgenic maize for the open market. Thirteen of 15 member states opposed the transgenic product because they felt there were not enough in-depth studies on the long-term effects of genetically engineered products as a whole. They also believed that the large scale use of *Bt* toxin raised environmental concerns and were troubled about the possibility of it increasing the use of pesticides.[11] In December of 1996, Denmark became the first European nation to require that all Maximizer® specified and other genetically engineered foods, generally, be labeled.

These European concerns over food safety and technology are greatly intensified by what many perceive as poor management, especially in Great Britain, of mad cow disease in 1996.[12] Related and perhaps exaggerated concerns have spilled over to transgenic crops. One focal point in the corn genome is the insertion of an ampicillin resistant marker gene.[13] This genetic combination, chosen to facilitate identification of recombinant seedlings, poses a small but hypothetical risk of spreading antibiotic resistance. In most cases the plasmid or carrier of the new gene bears only portions of the antibiotic resistance genes. Those gene segments that are present are only partially integrated into the corn genome and hence pose little risk of bacterial transfection. But in the case of the Novartis product known as CG00526-176, the entire antibiotic resistance molecule was attached.[14] In this case, a small but meaningful public health risk is almost certainly present.

Resolution

The labeling issue had sufficient force in Europe to lead to regulatory action. On July 31, 1997, "novel" food regulations came into place in all European Union member countries that require labeling. The specific language (article 8) of the Act requires genetically modified foods to be labeled if the modifications render a new food or food ingredient "no longer equivalent" to prior crop types, leaving considerable room for interpretation.[15]

EU member states apparently want more autonomy in the implementation of this decision. Some have challenged the European Commission to allow them to decide for themselves whether to allow the entry of unlabeled transgenic products into their country. So far the EU has granted only limited autonomy on the issue of labeling and importation. According to the European Union Directive, Article 16, member states have the right to veto a Commission decision to allow a product into market only if they believe that the product "constitutes a risk to human health or the environment."[16] In defiance of this narrow reading of the bill, Denmark has insisted that all unlabeled products be banned from importation. Many countries are using Article 16 to ban all *Bt* maize. Italy, Austria and Luxembourg banned transgenic corn imports, claiming that the antibiotic-resistant markers found in its production have not been sufficiently tested.[17] France has banned the product from being grown domestically but is permitting the corn to be imported if it carries special labeling. (Italy along with France produce 65% of the 33 million tons of corn grown in the European Union.)[18]

In mid-1997 Novartis made a major concession. In a break from the remaining industry members, it declared publicly that they *will* label their products. In doing so, they reasserted their claim that genetically engineered products are superior to conventional ones and indicated their interest in dispelling skepticism.[19] The politics behind this move are clear: Novartis' head Wolfgang Samo acknowledged that labeling was a proactive policy, not to win consumer acceptance, but to enhance farmers' willingness to use their seed.[20] The labeling will only be for two products, Maximizer® corn and *Bt* 11® corn seed. This labeling "victory" may be hollow, however, since Novartis plans to put labels solely on the seed bags and not the finished product.

In theory, Novartis hopes the message will tell farmers in the United States and Europe that they are receiving "added value" and quality by choosing to use the genetically enhanced seed. Downstream, European consumers will see labels stating only whether products "do," "do not," or "may" contain genetically modified materials.[21]

This decision was exactly what Dan Glickman, head of the USDA, had wished for. It also pleases U.S. industry, including grain traders and farmers, who opposed the initial EU decision calling for complete segregation of engineered versus non-engineered food products.

On July 31, 1997, food regulations governing genetically engineered food products came into force in all European Union member countries. The new regulations, particularly Article 8, state that genetically modified foods must be labeled if the new modifications render a novel food or food ingredient no longer equivalent to conventionally grown food.[22] Many people, environmentalists and scientists alike, are concerned about the interpretation of this law.

Labeling in the United States

Currently, the United States has no labeling requirements for genetically engineered foods. In May, 1992, the FDA declared engineered food products will be regulated no differently than foods created by conventional means. As a result, companies like Proctor & Gamble, Nabisco, Frito-Lay, Ross Labs, Pillsbury and Kraft do not have to apply labels which would otherwise alert the general public that the chips they are munching on likely contain soybean oil from genetically altered soybeans. When we called Frito-Lay, for example, and asked whether or not Fritos® contain oil from genetically altered soybeans, they answered us by saying, "We don't know, we don't test."[23] Most of the major food companies receive their pressed oil from the likes of large grain traders such as Cargill. U.S.-based Cargill is the world's largest grain trader, having a presence in 65 countries and providing grains to even more. Ross Labs, the makers of Isomil® and Similac® soy-based infant formulas on the other hand, admits to using engineered soybeans but has no disclosure statement on any of its products.[24]

The FDA Mandate

The FDA is guided by a decree called the Food Quality Protection Act (FQPA) of 1996. This act amended the Federal Food Drug and Cosmetic Act (FFDCA) of 1938. The new act was ratified in January 1997. The FQPA includes a section called "Consumer Right to Know" that promises within two years to include additional information on any new or unforeseen risks posed by pesticide chemical residues.

We are skeptical that this new rule will be applied uniformly to engineered and non-engineered crops. Monsanto has submitted to the Environmental Protection Agency (EPA) petitions for increasing the allowable residue levels for the chemical glyphosate, the active ingredient in Roundup®. As we showed, Monsanto successfully petitioned for the "safe, allowable" levels of glyphosate in soybeans to be changed from 6 to 20 parts per million prior to the introduction of Roundup Ready™ technology. What was considered "safe" for human consumption in 1987 was considered "safe" eight years later at three times the original tolerance. Will the new petition raise tolerances further?

Religious Concerns

For any of these reasons to be satisfactorily resolved we believe all transgenic research must be part of the public record. Our own experience has shown how far the public has been removed from the process of weighing risks versus benefits. When our organization attempted to get information about the apparent failure of genetically engineered cotton in Texas in 1996, not only did the request take almost six months, but the request was also sent directly to Monsanto. Monsanto itself was given the discretion to decide whether or not they would permit the information to go public. The government deferred to industry, even though the EPA requested that Monsanto supply an incident report of the failure before the product was used again in 1997. Instead of the report being treated as a public record, we were told that our request went to Monsanto for *their* distribution approval. When we eventually received the report, it contained no indication the product failed through any intrinsic flaw. Instead, Monsanto claimed that the product failed due to an unusually high year of bollworm pressure.

The FDA and major transgenic crop producers both argue that labeling is unnecessary in the absence of significant measurable differences in plant characteristics. Such a view ignores pressing moral issues for some consumers of genetically engineered crops. For many of the world religions, the derivation and purity of any given foodstuff carries theological importance. Many religious persons for whom diet is a significant part of their practice want assurances that the food they eat is not adulterated. Among orthodox Jews, the pure plant origin of a given product (such as a soy-based hamburger) is essential to avoid cross-contamination with any form of animal product, including animal derived genes, to assure that the product meets the conditions of *kashrut* or kosher law. Similarly, Buddhist vegans may want assurance that the food they are eating contains only plant-derived genetic material, since any animal form is considered sacred and inviolate. For these and related reasons, many religious practitioners have a bona fide claim on regulatory agencies to ensure that a given product does not contain additives or genetic material from nonkosher or animal sources.

A case in point is the position taken by Professor Ronald Epstein at San Francisco State University. Epstein points out that by Talmudic law, orthodox Jews may neither mix nor adulterate meat products with dairy products. Therefore, the origin of any oil or product used in preparing meat must be free of any dairy or non-meat additives. In theory, soybean or canola oils derived from genetically engineered plants may be considered adulterated and thereby nonkosher. Without a means of identifying a given oil as genetically engineered and kosher as specified in the Hebrew dietary codes, orthodox consumers may be prevented from learning the origins of their food products. Presently, all such foods are voluntarily inspected and labeled with the universal symbol for kosher, a U in a circle or have the "k" for kosher sign on them. A rabbinically trained inspector is required to be on the premises to ensure the use of kosher methods and materials in food preparation. Without suitable labels on genetically engineered products, the inspector may be prevented from doing his job.

Unlabeled foods also present a major moral dilemma for Muslims. Under the dietary rules embodied in the Koran, Muslims are constrained from eating certain foods (known as *Haram*). Those that are permitted are known as *Halal.* No plantstuffs are *Haram,* but products made from them, such as bread or pasta, may not contain any product derived from *Haram* animals, such as pigs, dogs, reptiles or amphibians. Bread made with animal fats may not be eaten, unless the animal was slaughtered by a Muslim in an appropriate manner. While it is presently unlikely that transgenic crops contain genes from a *Haram* species, some of the forbidden species such as certain bacteria or insects have been or may become gene donors. For religious reasons, it is critical that Muslims know whether or not the food they are eating is acceptable. Any residual uncertainty makes the food suspect, highlighting the importance of proper labeling: As with Kosher foods, *Halal* foods must be clearly distinguished from those which are *Haram*. The labeling of genetically engineered products is thus morally weighted for many consumers.

Biological reasons also dictate clear labeling. Where a product contains a novel gene, it may present previously unappreciated risks of allergenicity. This and related concerns are reflected in a 1997 pamphlet entitled, "Consumer Warning." Authored by Ronald Cummins of the Pure Food Campaign, this pamphlet alerts consumers to the possible allergenicity and pesticide residues that may be found in engineered plants. The simple policy of labeling would ease some consumer concerns by allowing them to avoid "contaminated" foodstuffs. The legitimacy of these concerns is underscored by the presence of a novel allergen in soybeans genetically altered with Brazil nut genes. According to the pamphleteer, consumers armed with this information could also alert their grocers and restaurants to boycott genetically engineered products.

Disclosure Duties

From an ethical viewpoint, disclosure of relevant information is critical in certain defined relationships. In the doctor-patient relationship, a complete disclosure of relevant information is deemed essential to the informed consent of the patient. However, state law, notably *Cobbs v. Grant* in California,[25] establishes a modification of the full disclosure rule. In Cobbs, a diabetic patient who was threatened with the loss of an infected leg was told, among other things, that the penicillin he was about to receive might produce anaphylactic shock and death. The patient declined the treatment and lost his leg. According to the Supreme Court ruling in the *Cobbs* case, the physician must not disclose so much information that he dissuades his patient from needed therapy. In this particular instance, the physician was not required to disclose remote possibilities of extreme ill effects, such as anaphylaxis from penicillin in a person with no family or personal history of allergies, especially when such disclosure might predispose the person to avoid medically needed intervention. By analogy, some industry proponents argue that disclosure of a distinction with (probably) little or no medical significance may unnecessarily dissuade the average consumer from eating a particularly nutritious food type. According to industry argument, disclosure of the existence of genetically engineered genes in food products would constitute such an unnecessary and potentially alarming disclosure.

The doctrine of "no distinction" was also used for recombinant DNA altered milk products. In this case, as we described in the Introduction, dairy companies were given a production boost by using recombinant bovine growth hormone or rBGH (also known as rBST, for bovine somata-tropic hormone). When milk producers supported by consumers balked at the usage of the hormone and sought to label their products as being rBGH-free, they were opposed by industry sources and limited by the FDA. As a result, a typical label (found on California's Stornetta Dairy milk) that declares on the front, "This milk does not contain the growth hormone rBST" is required to put an asterisk and a disclaimer on the back of the carton which declares, "Our opposition is not necessarily based on health issues. The FDA has tested rBST treated milk very rigorously and has found it to be identical to non rBST milk."

Disclosures of facts that are relevant to even a minority of consumers, such as the presence of trace amounts of phenylalanine in Nutrasweet® (an artificial sweetener) potentially ingested by persons with phenylketonuria, are nonetheless encouraged by the FDA. To many consumers, genetic engineering of a product is an important distinction that may determine their purchasing choice. A parallel example of such a rule is the disclosure of "organic" or "Certified Organic" on foods grown without pesticides.

A simple test of consistency is at issue here. A prohibition against labeling genetically engineered crops runs counter to the long-standing agricultural practice of permitting labeling for pesticide-free produce on certified organic products. Such labeling is permitted even though some conventionally grown products also contain little or no pesticide residues.

The requirements for what constitutes "organic" as determined by California Certified Organic Farmers include limitations on soil amendments, pesticide and oversprays of weed control agents. In 1998, the USDA will decide whether to include "genetically unaltered" as a requirement for its seal of organic approval. Any or all of these conditions may have special significance to a given consumer. By analogy, it may be critical for a householder to know if her produce is "uncontaminated" by a novel gene, even though for all but the most sensitive, the presence of that gene may have little or no biomedical significance.

Commentary

In the 1990s, case law and precedence have established that people in the United States have a "right to know" about the nature of the environment to which they are exposed. Food packages require certain nutritional labels; drugs and devices come with package inserts with full descriptions of risks and benefits; chemicals and commodities in states like California are labeled under Proposition 65 for their birth defect or cancer producing potential; and cigarettes carry a plethora of warnings. Today, virtually all articles in commerce carry identification and labels of one kind or another. For genetically engineered crops and their resultant byproducts to remain unlabeled flies in the face of both these precedents and common law practice. An industry position states that such labeling dilutes the force of truly necessary labels (e.g., "Poison!"). This position seems to patronize consumers by assuming that they are unable to make relevant distinctions.

Another industry argument is that according to projections, in the ensuing decade genetically engineered plants and by-products will be virtually omnipresent. To require all foods containing such products to be labeled will prove impractical and unreasonable and impose an undue burden on the producer. This argument fails to consider the fact that all foods are currently labeled, and that many of the labels carry information of no direct relevance to many consumers.

The requirement of disclosure is ultimately driven by society's perceived values. Where an issue has cultural value or merit, it is appropriate for it to be identified through appropriate disclosures. It may be acceptable to issue a disclaimer, concurrently, to offset any prejudicial effect of a label. As a way through this tangle of conflicting issues and values, it would be societally consistent to permit labeling of non-genetically engineered products. A disclaimer that "This product is free of genetically engineered products" would be comparable to Stornetta Dairy's claim of being rBST free.

We would predict that even such a simple disclosure would be opposed by industry. Such opposition reflects the position that flagging genetically engineered products permits them to be discriminated against and fuels irrational fears of the populace. Our view is if such fears are unjustified, let the manufacturer prove it: educate the public in appropriate forums about their safety. The prime reason for public fear about genetically engineered products is the industry's imposed secrecy about their development and identity. This angst is reinforced by industry's greatly accelerated production of transgenic varieties with little or no monitoring or pre-testing except to see if they "work" when doused with the appropriate pesticide. If the public had forcefully supported the initial forum for review, a biotechnology commission would have been created to discuss implications of transgenic crops. Had this commission done so, just as they did for cloning, little or no misapprehension about agrobiotechnology would likely exist today. But the fledgling biotechnology commission was disabled by corporate pressure. Congress abruptly disbanded the plans in 1995, and the industry has overridden each and every attempt at governmental regulation or labeling since. Such behavior fuels the very fears industry may now wish to subdue. The net result is an American public left much in the dark about genetic engineering and poised to openly revolt as activist groups make their case known.

---

[1] Angie Bailey, Pillsbury, Customer Response. Telephone interview. 18 February, 1997.

[2] M. Burrows, "Biotechnology's bounty," *New York Times,* May 21, 1997.

[3] R. Grossman, *Earth Island Journal,* 1997.

[4] J. A. Galloway, "rBGH still scares consumers," *Wisconsin State Journal,* January 26, 1996.

[5] Ibid.

[6] Reuters, April 2, 1997.

[7] J. Fox, "Biotech food labelling: polite hedging or loggerheads?" *Nature Biotechnology,* Vol. 15, April 1997, p. 308.

[8] D. Butler, *Nature* 384. 28 November, 1996.

[9] A. Abbott, "Austrian gene food petition puts pressure on European partners," *Nature* 386: 345, 24 April, 1997.

[10] Reuters News Service. Zurich, Switzerland. March 20, 1997.

[11] European Parliament Resolution on GMO Maize, Council Directive 90/220/EEC, April 8, 1997.

[12] Buckley, Neil. "Call for labeling of gene modified food," *Financial Times.* 25 July, 1997.

[13] R. Hoyle, "Novartis' new labeling policy creates confusion," *Nature Biotechnology,* 15: 1997.

[14] J. Hodgson, *Nature Biotechnology,* 15: 1997, p. 310.

[15] Business and Regulatory News, "EC food regulations," *Nature Biotechnology,* 15: 1997.

[16] Article 16, European Union Directive.

[17] Associated Press, Strasbourg, France, April 8, 1997.

[18] L. Gale, Greenpeace International. "Italy joins opposition to GMO maize." 7 March, 1997.

[19] B. Feder, "Biotechnology company to join those urging labels on genetically altered products," *New York Times,* 24 February, 1997.

[20] R. Hoyle, "Novartis' new labeling policy creates confusion," *Nature Biotechnology,* p. 395.

[21] Buckley, Neil. "Call for labelling of gene modified food," *Financial Times.* 25 July, 1997.

[22] *Business and Regulatory News.* EC food regulations. *Nature Biotechnology,* 15:1997.

[23] Theresa Broadbent, Frito-Lay nutritionist. Telephone interview. 6 June, 1997. At a later date (10 February, 1998) Ms. Broadbent acknowledged Frito-Lay did use genetically engineered corn and soybeans and "planned to increase its use in the future."

[24] Ross Labs customer service. Telephone interview. 4 June, 1997.

[25] *Cobbs v. Grant,* 502 p. 201, 10 (1972).

11

Recommendations

By our reckoning, the burgeoning use of transgenic food crops constitutes a nonconsensual experiment on a mass scale. For the first time since the Amendments to the Food and Drug Act mandated disclosure of food ingredients, people are being asked to accept and ingest foodstuffs without being told when and if their food contains engineered gene products. While we continue to acknowledge that for most people direct health risks are likely to be small, for others the risks may be significant. Other issues are not so easily dismissed. Large numbers of persons who believe their religious or moral beliefs are threatened by consuming engineered foods clearly have reasons to object to the veil of ignorance created by nondisclosure. At a minimum, we all need to be told what we are eating.

At another level, we are concerned about the sheer magnitude of the transformation of basic staples and the impact this transformation may have on ecosystems and human well-being. While many of the consequences of such an expansion—such as disease susceptibility or microecosystem disturbance—remain largely hypothetical, we believe even the minimum requirements for responsible action to anticipate adverse consequences from transgenic crops are being sidestepped. We have seen concerns of regulatory agencies like the EPA and USDA/APHIS swept aside by assurances that the novelty of transgenic crops is exclusively limited to their growing conditions and enhanced pest resistance. Further assurances of total safety from industry scientists have convinced the FDA to downgrade concerns about public health to a situation of watchful waiting. We believe the clean bill of health being offered across the board for transgenic crops is premature and ill advised.

Many of the herbicides for which safety is being assured, notably glufosinate (Liberty™) and bromoxynil (Buctril™) are incompletely studied and carry residual risks from even modest contamination above present tolerance levels. We do not believe long-term safety for either has been scientifically proven or accepted. Other herbicides, like glyphosate, appear relatively safe using the older tolerances established by the EPA, but we believe it is unwise to raise tolerances simply to permit more widespread use of transgenic technology. Similarly, we remain unconvinced by the data used to prove that *Bt* engineered corn is safe, especially when widely consumed by cattle or people. This is because the toxoid levels needed to control corn borers and other pests is almost certainly going to increase (as it has in cotton crops), putting more of this protein into the final product. The permeation of the marketplace with *Bt* products (in cotton, corn and soybeans) has greatly increased the likelihood of selecting for resistant strains of Lepidoptera species. Should such resistance occur on a mass scale, integrated pest management programs will be sorely affected. We therefore recommend a wholesale review of the regulation, testing, and inspection of all engineered crops be undertaken.

As a second recommendation, we urge a moratorium on usage of *Bt* crops until appropriate studies have been done to assure that no *Bt* resistant strains will arise, and until chronic feeding studies have been completed to assure a reasonable confidence in the safety of the toxoid-containing products.

To offset the prospect of these and other potential effects, we recommend the establishment of an Agricultural Biotechnology Commission. We are especially concerned by the shortsightedness of governmental agencies. When we asked the deputy director of the USDA's Biotechnology and Scientific Services, Mr. Arnold Foudin, if the USDA had a committee to look at the long term effects of transgenic crops on ecosystems, he responded "No, in fact most of the effects will occur almost immediately." The Commission should be expressly charged with examining the ecological, health and safety, and any other long-term consequences of the widespread introduction of genetically engineered foods. Commission members should also be asked to draft a statement of ethics itemizing the elements of corporate responsibility that need to accompany large scale introduction of transgenic crops.

We would also charge APHIS with the responsibility of setting up a system of oversight and monitoring. At the core of such a system would be a formal tracking mechanism to ensure at least a statistically meaningful subset of transgenic crops entering the marketplace be closely followed, and that consumers who are at the end of the transgenic food chain be monitored for possible adverse effects. If such a requirement were in place during the 1970s and 1980s when genetically engineered L-tryptophan was first marketed as a "safe" food supplement, the 37 people who died and 1,500 or so who were adversely affected by a fermentation contaminant and went on to develop eosinophilia myalgia (a painful and debilitating circulatory disorder) might have been detected earlier.

A third element of public oversight and input is needed to encourage multinational corporations like Monsanto to stay with their assurances to "feed the world." A consolidated objective for such companies is the creation of crops requiring fewer chemicals with less dependency on fertilizer, herbicides and insecticides and that bear nutritional advantages to the end use population. Specific examples include increasing amounts of critical amino acids like cysteine and methionine that limit how much human protein can be made from plants; incorporation of enzymes like phytase to reduce phosphate levels; and increasing plant tolerance to drought and reduced soil nutrients.[1] Achieving these developments is particularly crucial for developing countries facing chronic protein and calorie malnutrition and poor quality protein sources.

While a few companies are developing products that are moving in these directions, the majority of the biotechnology infrastructure is presently focused on high economic returns for their investors or privately held company stock holders. A few companies, notably Agracetus, Inc., is using transgenics to make potentially valuable medical commodities. They are marketing products which express therapeutic proteins in genetically engineered plant cell cultures and corn and soybean leaf and seed. But Agracetus and a few others are going against the grain of present trends in the biotechnology sector. The most forceful voices, notably companies that have been in the agricultural chemical business, have urged greater investments into agrobiotechnology and expressly disapproved pharmaceuticals.[2] The proponents of this view note that in Europe, in particular, the pharmaceutical industry comprises a market of less than 0.02% of the Gross National Product. In contrast, countries like the Netherlands have 20% of consumer spending and 18% of the workforce concentrated in the agricultural food sector. We think equally valid arguments in favor of pharmaceutical investments can be made, for instance, in encoding vaccine components in animal plant biomass. The market for hepatitis B vaccine is in the hundreds of millions of dollars.

As long as we continue to allow corporations to develop crops that in the end will economically benefit only them, we will never see the full potential of a science that could be used in much more beneficial ways. The U.S government is not guiding what is to be developed; they are simply sitting back and offering approval for the corporate products. In the words of Arnold Foudin, "the U.S. government does not dictate to industry." As long as this is so, we are at the mercy of corporate decision making and the resulting short-term, high-gain product mentality.

Fourth, for biotechnology to make a significant impact on the food balance sheet of the world, it must consider the present conditions of the conversion of valuable plant food into animal protein. The demand for animal protein is expected to escalate sharply as more and more developing countries follow the lead of their affluent neighbors. Presently, the success of plant to animal conversion in such countries is 8.5 pounds of plant protein for every pound of animal protein. In the more highly mechanized and efficient production facilities common in many developed countries (e.g., for chicken farming) the plant to animal ratio is closer to 2.5:1. This observation suggests that more protein is wasted in developing countries who import European or American products than when such products are used or processed at home. The exportation of finished protein supplements in the form of soybean curd (tofu), corn meal, or bread intended directly for human consumption is more cost efficient in terms of protein and calories than exporting transgenic seed crops intended for animal use. Given these differences, we would recommend a policy of exporting finished products such as corn or soybean meal or wheat flour directly intended for human consumption rather than for animal feed.

Fifth, we would like to see much more independent research on the possible health consequences of ingesting transgenic proteins like those made by the genes for glufosinate resistance or *Bt* production. None of these genes are encountered in nature in the same or modified forms. But they will be consumed more regularly and in higher quantity than ever before if they are expressed in transgenic crops that are projected to comprise 50% or more of the market. Specific research is needed on *Bt* toxin to insure that its ingestion is not associated with allergenic responses or detrimental effects on the normal gut flora of vegetarians or others who consume vegetable protein as a large proportion of their diet.

As our sixth recommendation, we of course would like to see labeling of transgenic products made mandatory in keeping with the ethical maxim of full disclosure. The public deserves greater access and control of the critical data concerning transgenic status, if for no other reason than that they are the ultimate consumers of the finished engineered products. For every other foodstuff they now have the right to knowledge of nutritional content and ingredients—they deserve no less for transgenic products. We remain unconvinced, however, that such a step would be the most sufficient response to controlling the transgenic conversion of our foodstuffs.

Our view is that it is a smokescreen for larger issues of distributional justice—who gets to make and keep transgenic seed, for example, and the question of retention of germ plasm resources in developing countries. If the industry continues its rapid conversion of our foodstuffs to transgenic, the labels will be a moot point because all of our foods will contain engineered organisms. The larger issue is the necessity of maintaining genetic diversity. Towards this end, we would urgently recommend setting aside a percentage of global crop acreage to house non-engineered seeds.

We are disturbed by the present forces motivating transgenic seed development: the motives for product expansion appear to be centered on products that carry high technology fees, that rely on parent company herbicides, and that afford limited benefits to consumers while increasing profit margins through reduced work forces on farms. The demise of the small farm is currently being offset by a modest expansion in organic farming methods. Traditional organic techniques rely on the efficacy of *Bt* spores and treatments reducing crop predation by insects. By selecting for resistant organisms, the present strategy to incorporate *Bt* toxoid in plant tissues will almost certainly decrease the utility of this organic crop control method and put further pressure on the small farmer to shift towards transgenic crops. If the present steps to incorporate language encouraging labels of "organic" for most transgenic crops are successful, it is likely that still fewer competitive advantages will remain for small organic farmers. This is so because the megasize farming operations employing *Bt* derived products could be largely pesticide free, even though they contain a novel gene product. We therefore recommend that the label "organic" exclude *Bt* containing crops created by transgenic engineering, and that *Bt* be considered a controlled substance by the EPA whose pattern of use or application is subject to strict governmental oversight.

Finally, we fundamentally disagree with those who maintain that transgenic crops are no less organic than are traditional hybrid or other selected varieties. By virtue of their humanly manipulated genes and their potentially destabilized genomes, these crops are both novel and potentially disrupting to the natural balance of nature. We have already documented how widely dispersed some transgenic plant pollen can spread, and agree with the authors of a recent review of risks from transgenic crops that every genetically modified organism should be evaluated for its ecological risks on a case by case basis.[3]

Presently, no mechanism exists to ensure such review. For this reason, we propose as our seventh recommendation, the expansion of the present requirements for Environmental Impact Studies to include every sub-type of transgenic commodity, with environmental monitoring continuing on an ongoing basis for at least two years on at least one plot of each planted variety before widespread application. Without such review and tracking, any ecological dislocation that might occur will be lost in the hodgepodge mix of weeds and secondary growth that follows at the periphery of any agricultural development.

We are also concerned about the absence of transgenic safety corridors lining major field applications. For transgenic plants that have natural congeners or related species, we believe at least some hedgerows and peripheral ecosystems should be monitored for at least 2–3 growing seasons to assure that transgenes do not escape to infect native plants. This problem is especially troublesome as herbicide applications move from the farm to the logging enterprises around the world. With such expansion of use, it is possible that the novel genotypes introduced, say by glufosinate resistance, will suddenly become selectively advantageous. Should this happen, a pollinated feral species adjacent to a logging operation might be given a selective advantage in a herbicide sprayed forest. Should such an event happen, the movement of transgene-containing species from farm to ecosystem will occur in a manner similar to the movement of antibiotic resistant bacteria from the hospital to the community. We are neither prepared nor equipped to deal with this eventuality. Affirming the safety of this powerful new technology would appear to demand nothing less.

Comment

At present we are proceeding as if there are no downside costs to the massive introduction of genetically engineered plants. This shortsightedness is reminiscent of the early days of the antibiotic revolution in which penicillin and other first generation antibiotics were squandered for trivial uses, leading to the expansion of antibiotic-resistant strains in nature, feedlots, and ultimately people. Among these trivial uses were the application of penicillin in pear orchards to control rust and fungal growth. We are doing something strikingly analogous with the indiscriminate engineering of *Bt* genes into cotton and other nonessential commodities. What will be the impact of such overuse? Have we factored the demise of *Bt* sensitive target organisms into our agricultural planning? While the public sector has not undertaken such a study, it is clear that industry has. Industry sources we have spoken with acknowledge that their use practices of *Bt* will make this technology obsolete within 7–10 years. These corporations are ready to move on the next genetic modification—only this time it will not mimic a natural pathogen, but will likely once again resurrect a chemical solution for weed control.

If we are to offset this wanton disregard of environmental ethics, we need a means of regulatory oversight at the Congressional level. To do otherwise is to face the tragic scenario now eclipsing the antibiotic era: too little concern for second order consequences plus overreliance on critical chemical means of control led to their demise. We face the same disaster with genetically engineered herbicide tolerant crops in the near future. Can we rely on corporate ingenuity to develop new alternatives? We think this is a treadmill all of us would like to avoid. To do so means scaling back reliance on monolithic chemical control and returning to more integrated pest management strategies.

Greater reliance on organically grown foods may be the answer to much of the concern about chemical pollution. But the present trend towards greater rather than lesser herbicide reliance brought about by selective herbicide tolerant crops—albeit for a small cross section of selective chemicals—is going in exactly the opposite direction. We would do well to examine this trend now, and to design strategies to reduce rather than increase chemical dependency by thoughtful design of the next generation of transgenic crops.

In our opinion, future "experts" who look at our crash programs for allowing one herbicide like glyphosate to dominate agriculture in the late 1990s will be astounded at our hubris and ignorance. There is little doubt that this herbicide has fewer visible side effects and safety problems in terms of acute toxicity than many others. But its saturation of the environment is almost certainly going to have adverse ecological effects, especially at the microecosystem level, effects with which a future generation will have to cope.

At this time, it would be unwise to brush aside concerns about genetically engineering our food supply in favor of touting its benefits. For example, we do not know how the *Bt* protein ingested in such large amounts will act in the digestive system of humans after sustained use. In a classic example of understatement, a USDA/APHIS official admitted that the potential human reaction to the ingestion of *Bt* has "not been studied to death."[4] To our knowledge, it has not been studied at all. Cautious regulation would allow us the opportunity to avoid these potential risks while insuring that biotechnology contributes to our social and long term environmental well-being.

---

[1] See J. F. Burke and S. M. Thomas, "Agriculture is biotechnology's future in Europe," *Nature Biotechnology* 15: 695–696, 1997.

[2] Ibid.

[3] Ellstrand, N. C. "Pollen as a vehicle for the escape of engineered genes," Biology of Risk Assessment; S30–S32.

[4] Arnold Foudin, Deputy Director of APHIS/USDA Biotechnology Scientific Services. Telephone interview. 6 August, 1997.

12

Conclusion: Against the Grain

The Big Picture

In documenting the extent of conversion of foodstuffs to genetically engineered varieties, we have been struck by the arbitrary nature of the choices of genes and products corporations have brought to market. In virtually all instances, short-range economic considerations have driven the selection of genetic products rather than choices based on long-term objectives or public benefits. Instead, we have found companies favoring transgenic manipulations to enhance the value of their own patents, notably those on bromoxynil or glyphosate (Buctril® and Roundup®). As we have seen, companies such as DuPont and Monsanto selectively concentrate on moving into seeds those genes that confer resistance to their own herbicides. Given the amount of research into congeners of these herbicides, it is highly likely such companies would have found new varieties of these herbicides with better activity/safety ratios than glufosinate, bromoxynil, or even glyphosate. But by transferring genes for selected herbicide tolerance into food crop plants, such companies have committed a whole industry to their original products, and walled off new discoveries from the marketplace.

Once committed to a particular product, we have seen the corporate need for rapid commercialization drive the registration, review and marketing process of the start-up genetically modified product, reducing any chance of incorporating newer and safer variants. The extraordinarily costly development process, which includes synthesis, production and registration, takes from 5–7 years, ensuring that once a corporation is committed to a given chemical entity, its options for shifting priorities become highly constrained.

This process is unfortunate on two counts: first, because it condemns the farmer, consumer and producer alike to an often imperfect chemical choice. Second, rapid commercialization carries an opportunity cost, limiting the resources that might have been committed to a better product line. This limitation has proven especially unfortunate for Rhône-Poulenc's bromoxynil, an herbicide with teratogenic and carcinogenic activity. Few public health officials would have recommended full-scale development of this toxic herbicide in transgenic crops like cotton. More to the point, the existence of the novel toxic metabolite, DBHA, produced in all transgenically transformed BXN® cotton plants, should have doubled the registrant's burden. Instead Rhône-Poulenc has avoided having to do any but the most basic toxicity testing of DBHA by relying upon data from bromoxynil's prior successful registration. This avoidance runs contrary to the fact that the unique genetic attribute of the BXN® cotton plant converts bromoxynil almost entirely into DBHA. Residues for bromoxynil in BXN® type resistant transgenic crops are thus qualitatively different from the non-engineered varieties previously studied. In our view, this fact alone warrants a new review of bromoxynil tolerant transgenic crops.

This point appears to be lost on the EPA which is treating both bromoxynil and DBHA as comparable chemicals in terms of toxicity. The critical reality of higher than expected residues of DBHA in transgenic cotton should give the agency pause. Instead, EPA officials seem to be putting aside any toxicological risk assessment of the transgenic crop, as if bromoxynil is the sole residue in question.

In the grand scheme of things, we would like to see the EPA have the authority not merely to regulate what is given to them after it is developed but to advise and guide research energies towards more environmentally benign and healthy alternatives. These alternatives include the "no herbicide" options of new cultivation techniques, integrated pest management strategies, and more benevolent chemicals such as the pyrethrins. Instead the corporate producers of patented transgenic products increasingly appear to dictate to the grower and consumer a more circumscribed number of choices for production seed.

Corporate Imperatives

With the advent of other transgenic crops, we will undoubtedly see the commercialization of genetic technologies that further limit freedom of choice, both for consumers (who will not be told about the differences) and growers (who will not be able to select among competitive varieties of crop seed). The ultimate dilemma for consumers is having to ingest products without a label which informs them of the nature of the purchased product. This omission will be reinforced by current legislation and policies which restrict states from instituting any kind of informational labeling, unless such labeling is strictly linked to health outcomes. We find this omission symptomatic of a broader neglect of consumer education about genetic engineering. Suppression of knowledge also limits debate, and ultimately increases consumer resistance to change.

We are also concerned by industry-led discouragement of any broad scientific oversight of transgenic agricultural products. Led by Monsanto, one corporation after another has asked for exemption of their innovations from further review or tracking. As a result, we lack knowledge about many critical aspects of this new technology: the long-term fate of genetically engineered crops and their genes, their potentially toxic byproducts; and the ultimate impact of committing more and more of our productive land to animal feed crops with a limited genetic base. In giving the green light to genetic technology, we may be consigning the planet's food production into the hands of a few corporate monopolies whose interest in the bottom line may transcend their commitment to serve the public interest.

Unless and until there is reform of such head-in-the-sand exemption policies, we are likely to be condemned to blindly embracing genetically engineered foodstuffs. Lacking the means of identifying these products or their consumers, all future events stemming from hidden risks (as occurred with fermentation produced L-tryptophan and eosinophilia myalgia) may be missed. Consumers who might face increased risks of cancer from excessive bromoxynil levels or ill-health from consumption of *Bt* toxoid containing foodstuffs will neither be able to be tracked nor found.

We have identified a number of critical issues raised by genetically engineered crops, issues which distinguish agricultural biotechnology from the more traditional agricultural practices. Certainly "standard" agriculture already displaces native species and diminishes the diversity of natural stocks by selecting only a few cultivars for intensive use. The widespread use of genetically engineered plants not only aggravates these already unfortunate trends, it also adds a new level of chemical dependency and further narrows diversity to those few gene lines that are "successfully" engineered. Corporate values inevitably lead to selection of value-added crops with great economic benefit to their customer growers, but little or no enhanced nutritional value to the consumer.

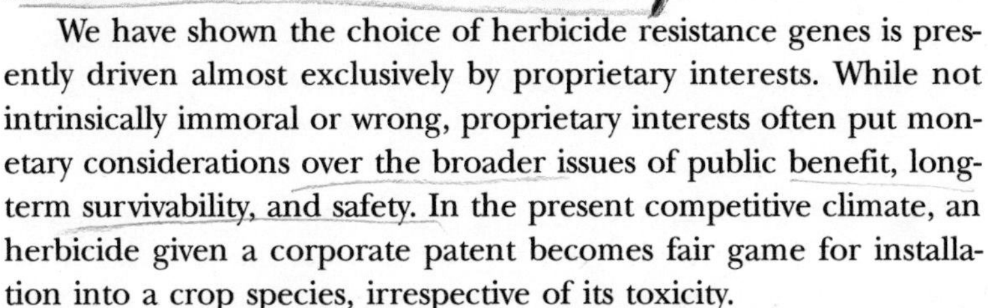

We have shown the choice of herbicide resistance genes is presently driven almost exclusively by proprietary interests. While not intrinsically immoral or wrong, proprietary interests often put monetary considerations over the broader issues of public benefit, long-term survivability, and safety. In the present competitive climate, an herbicide given a corporate patent becomes fair game for installation into a crop species, irrespective of its toxicity.

The reasonableness of this genetic choice is dubious for two reasons. First, the choice of a resistance gene is not driven by safety or utility as much as it is by direct commercial motives. Second, the germ plasm of the crop, once genetically engineered, becomes permanently tainted by an often non-native (i.e., not species specific) gene. In some instances, the germ line of the seedstock is permanently transformed. The full range of long-term consequences of these new gene insertions is presently unknown. Certainly, the new protein product of the introduced gene, usually an enzyme that detoxifies the herbicide or its metabolites, presents an additional component of the plant which may alter the bioavailability of essential nutrients or becomes a toxic or quasi-toxic component of the plant itself.

A further concern is that the gene product survives in the kernel or bean of the plant, adding unspecified amino acids and, possibly, new toxic byproducts to the commodity. While the risks of such additional proteinaceous materials are largely unknown, they certainly deserve multigenerational testing in animals before food or animal feed is supplemented with genetically engineered product. Where such evaluation was omitted, for instance, in the production of L-tryptophan, new and highly toxic byproducts were found to be contaminants.

In the instance of genetically engineered products, we already know of the potentially lethal allergenic consequences of adding methionine-rich Brazil nut proteins to soy products. In this instance, the production of a highly allergenic protein in the seed was fortuitously detected and averted. But the possibility that similar errors could arise from insertion of other new genes remains unexamined, and downplayed by industry assurances that new proteins are genetically not of the type known to be allergenic.

The secondary effects of genetically engineered crops must also be considered. Ecological spread of "advantageous" genes put into crop plants remains a serious possibility. This risk was highlighted by the spread of transgenically introduced herbicide-resistance genes from a genetically engineered canola plant into a nearby weedy species (field mustard) and the widespread dispersion of transgenic pollen.[1] It is also conceivable that the "marker" genes that confer antibiotic resistance onto the parent plant (used to assist researchers who must find the successful transgenic plant in the early stages of propagation) may also spread into natural ecosystems. While of relatively little importance in light of the vast numbers of antibiotic resistant bacteria already present, the possible unique selective advantage conferred onto non-native species by such an introduction must be seriously considered.

Ultimate Issues

The ultimate question posed by genetically engineered crops is very much akin to the one raised early in the genetic revolution: Do we have the wisdom "to play God"? Or put into more contemporary language, do we have the foresight and intelligence to substitute human selection for natural selection? At one time, questions of this sort were largely academic. But with the enormous increase of production of genetically engineered crops and the scale-up of production to occupy 50 percent or more of certain key grain crops, we have to ask if we are not backing ourselves into an evolutionary corner in which the dominant plant types, certainly on the North American continent, are human engineered transgenic species. What will be the consequence of millions of acres of such plants on reduced diversity and long-range adaptability?

So many of the genetic choices made by the current generation of transgenic engineers appear to be short-range that we are concerned that mid-range (5–15 years) plans to anticipate human needs in the beginning of the 21st century are being ignored. Given the global climatic changes that increase the risks of desertification and shift growing regions, should we not be concentrating on selection of plant types with maximal range and drought tolerance? While such features are presently included in the summaries of the plant characteristics sought by a few small transgenic producers, there is no evidence that seed companies are actually planning for long-range changes such as those that may be brought on by global warming. For that matter, there appear to be no efforts of transgenic companies to tap into the reservoir of genetic types carried in seed germ plasm banks, or to conserve vital germ plasm to anticipate such changes. This is especially unfortunate since transgenesis may afford the only means of genetic change fast enough to ensure new plant types keep up with the accelerating rate of planetary change.

The net result of our current patterns of choice is to permit corporate values to dictate the decisions about what plant species will dominate the agricultural regions of the planet. These decisions are being driven by short-term interests and not by those of the indigenous peoples or local farmers. We question whether such corporations have either the wisdom or the prerogative to make such choices. Moreover, such dominance carries an enormous "opportunity cost." By committing large growers and governments to crops with genetically programmed chemical needs or dependencies, we are creating a chemically fixed biosphere in which specific herbicides are a feature of foliar type or seed pattern.

Against the Grain

To some, these concerns may appear to be nothing more than the carping of some anti-technology group, or, worse, simply a matter of taste. Why are transgenic crops any more or less detestable than were the hybrid types created decades ago that so dominate North American agriculture? The answer in part is that both are abhorrent to the extent that they fix and delimit the evolutionary potential of the seed crop at issue. Fortunately, to propagate hybrids two or more parental lines that carry the features that make for hybrid vigor or heterosis must be maintained. Should the farmer wish to move on with a new crop, in the past he could plant the F_2 generation seed. Indigenous peoples also had the option to begin again the laborious effort to select a new and desirable landrace through introgression of new genes.

But a transgenically engineered seed "belongs" to the company. Its genetics are fixed. Each year, the farmer must contract to plant his seed again from an ever shorter list of seed providers and an ever-shrinking number of crop varieties. Increasingly, the seed sellers are nothing more than an extension of a large chemical company conglomerate which also makes the herbicides it wants to sell to the farmer. Even if the farmer replanted his seed from a transgenic crop, he would of necessity have at least a partially herbicide-dependent crop with the same gene product in it (a result of self-fertilization). Hence, the use of transgenesis virtually assures the continued dominance of chemical companies in agriculture and the resulting co-dependence of large growers on the economic benefits offered by their continued reliance on a given herbicide or insect-resistant gene.

We strongly believe in this sense that all transgenics go "against the grain" of the patterns and rules of a natural world. They have characteristics that are chemically, not naturally, driven. Their very success drastically limits the diversity of genome types for any given food crop. And they carry genes that may confer selective advantages to weedy species. Ultimately, we question whether transgenic foods are in the consumer's best interest. Any consumer preference for "natural" foods (i.e., non-transgenic crops) is presently honored in the breech. No requirement exists for the manufacturer to disclose the makeup of its product, so no one can be sure the products they are consuming are organic, transgenic, conventional, or strictly kosher.

This omission in itself raises the ethical issue of the right to know. We have argued earlier in this book that this right has been widely recognized as a universal principle in occupational health and in food labeling generally. Not every labeling requirement discloses information deemed relevant to all consumers. Nor will the data about the presence of transgenic content. But it will be useful for those who wish to know, for religious, political, moral, health, or even emotional reasons, the origins and makeup of their food. To simply declare, as does the USDA, that a transgenic crop designation is relevant to farmers but not to consumers (the FDA's position) is unacceptable. Clearly, both traditional and organic farmers *and* consumers want this information. We believe that they should get it.

The larger issue is the degree to which transgenic commercialization of feed crops will lead to more and more acreage being planted with corn or soybeans intended for animal instead of human consumption. Presently, of the 8–10 million acres planted with soybeans, 95% of the crop, in the form of meal, is intended for animal feed, with 5% of the soy meal going to humans. While 80% of the oil is processed for human consumption, virtually all of the protein is being poured down the bellies of our livestock. As Frances Moore Lappé wrote in her seminal book, *Diet for a Small Planet,* (Ballantine Books, 1975), the typical 9–10 % conversion rate by which animals in developing countries turn plant protein into animal protein dissipates almost 90% of the food value of the crop.

Further, dependence on animal protein goes together with dependence on synthetic fertilizers. About one-third of all of the protein in the world's diet comes from synthetic nitrogen fertilizer. Much of this dependency comes from the intense pressure to grow feed crops—food for animals—not food for people. With increasing economies of scale and a growing livestock base, transgenic technology aims at reinforcing this usage. We have already shown how the EPA has modified its residue tolerances to permit greater forage use of transgenic crops. Use of herbicide tolerant crops virtually guarantees that beef, poultry and pork will have higher contamination levels of selected pesticides than such livestock had previously. Special tolerances for Roundup® herbicide residues in silage were instituted to increase the utility of Roundup Ready™ soybeans in animal feed crops. With the bioconcentration factor that operates for many of these herbicides, we will be exposing people, especially children and infants for whom soy is a larger component of their diet, to a largely untested and different burden of pesticides than before.

The use of transgenic crops on the massive scale now contemplated is above all a vast experiment in agronomy. Its very size and scope assures that millions of people will also become unwitting subjects in a mass experiment. To say otherwise is to misconstrue the scope of this new enterprise. For the first time, corporations are exercising direct dietary control over the genetic makeup of whole orders of foodstuffs on the planet. In our ultimate hubris, we are not simply creating new plant varieties and watchfully waiting. Instead, like Zeus consuming his progeny, we are eating our own genetic experiments, putting our children and our children's children in possible harm's way

In doing so, we may be allowing novel gene sequences, like those for *Bt*, to enter our own microbial microecosystem, colonizing gut microorganisms and ultimately returning those genes to the ecosystem generally. We already know these gut bacteria break down the residues of herbicides differently than do our own enzyme detoxification systems.[1] How will they handle novel gene products?

What makes us think our new gene choices are any more benevolent than were our older ones? Through blind genetic selection, we developed hundreds of forms of cassava, only to discover that each bore fruits with varying amounts of poisonous cyanide; we selected for dozens of sweet potato varieties contaminated with varying amounts of "natural" progesterones; and we bred cats, cattle, and sheep strains contaminated with viruses and prions and other potential human pathogens. Yet, we blithely believe our creation will be beneficent.

This sublime misconception fuels the present genetic revolution of food crops and threatens to undercut its potential utility. In the long run, by neglecting the evolutionary consequences of our immediate preferences we potentially commit future generations to a world dominated by late 20th century values of the "good." Is Roundup® the "best of all" herbicides? We certainly do not profess to know, but does anyone else? We are committing a whole industry to this one chemical.

The other step we are now taking, channeling the remaining genetic resources in our food crops into a few genetically engineered sub-types, may circumscribe our future options, perhaps irrevocably. The step from domestication, in which virtually all of the native gene stocks are simply re-assorted into varying sub-types or breeds, to the fixation of a subset of genes into a single commodity and its variants is a huge departure. Anyone who has bred dogs has observed how quickly the prototypic canine wolf-like form will re-emerge from outcrosses between distant breeds. This can happen because most of the genes for "wolf-ness" are retained in the plethora of breeds that make up the annual Westminster Dog Show. But in plants and food crops we are now confining the germ plasm to a few archetypal "productive" types with five or six desirable characteristics. When we fix those types with the most desirable characters with supernumerary pest-resistance or herbicide tolerance genes, we virtually guarantee the diminution of options for future recovery of gene stocks that may have contained essential and unrecognized genotypes. When we "outcross" transgenic crops, the trans-gene goes along with any fertilization. But this distant concern about loss of diversity, a view expressed ably by many others, is overshadowed by more pressing issues: We lack a forum to participate in public dialogue and choice. Corporations are understandably mum about their proprietary developments and their problems. No labels announce a food product's transgenic origins, locking out the consumer from participating in meaningful decisions.

But perhaps this premature enterprise will have a proverbial happy ending. If we discover early enough that it is a mistake to put bromoxynil tolerant genes and ultimately bromoxynil itself into cotton plants that supply us with oil and clothing, we may be able to stem the tide of any harmful aftermath. If we are able to count the number of newly deformed or cancerous children in the next generation, and make adjustments accordingly, just as we did when we proscribed the introduction of carcinogenic hormones like DES into our cattle as growth supplements, we may stave off disaster. But the worst case scenario of all is if we allow corporations to thwart the rights of consumers to know—and epidemiologists to track—the genetic footprints of their potential folly. The ultimate foolhardiness is if we bow to such pressures and fail to label and track this new generation of genetically adulterated products. Certainly, if we have learned anything, it is that our hubris in dominating nature often puts us into harm's way. If we fail to mark the path with suitable disclosures and labels, and instead allow it to be littered indiscriminately with natural and transgenic seeds alike, we may discover that, like Hansel and Gretel, we are unable to find our way out of the woods.

This admitted hyperbole may become a reality if we do in fact allow most or all of our next generation of food crops to be limited to genetically engineered types. By further reducing the germ plasm available to our farmers, we run the risk of creating an evolutionary endgame. This debacle may indeed spell the end of the grain as we have known it: from a diverse group of strains and cultivars representing the cultural richness of our past, we will be substituting a small number of genetically engineered subtypes geared to modern agricultural technologies. Some might object to the cultural relativism implied by this criticism by noting that the new crops are simply adjusted to reflect the post-modern technoculture of crop raising. To them we simply say, diversity, be it in cultural types or agricultural practices, provides a hidden guarantee against evolutionary disaster, ecological disruption, or simple crop failure. By ensuring that seed crops will be selected by the persons who grow them and not by the corporate detail person, we will assure the perpetuation of the deeper linkages that bond the farmer to the soil, and ultimately, the consumer to the natural earth on which we all depend.

---

[1] Kling, James, "Could Transgenic Supercrops One Day Breed Superweeds?" *Science,* 274: 180–181.

[2] Delcensarie, R. *et al.* "Syndrome pseudo-appendiculaire tardif après intoxication au Roundup®. 22 *Gastroenterology Clinical Biology* 21 (1997):435–437.

Glossary

A

Adaptation: the process by which an organism undergoes modification so that its functions are more suited to its environment and its changes.

Agar: a seaweed derived medium for growing plant embryos or other organisms in tissue culture.

Agbiotech: short for agricultural biotechnology; the organized application of genetic manipulations to plants.

Agrobacterium tumefaciens: a bacterium which causes crown gall disease in a range of dicotyledenous plants, especially coastal member of the genus *Pinus.* The bacterium can enter dead or broken plant cells in a living organism and transfer a tumor-inducing portion of DNA in the form of a plasmid. The plasmid then integrates into the plant's own genetic material, constituting a natural form of genetic engineering. Strains of *A. tumefaciens* can be artificially engineered to introduce selected foreign genes of choice into plant cells. By growing the infected cells in tissue culture, whole plants can be regenerated in which every cell carries the foreign gene.

Agronomist: a scientist who studies agriculture and its systems; cf. agronomy.

Allele: from "allelomorph" meaning one of a series of possible, alternative forms of a given gene that differ in DNA sequence but produce a similar product, for instance a blood group or plant protein.

Amino acid: a subunit of a protein, usually comprised of a short sequence of carbon atoms or ring structures with an amine group on one end and a carboxyl group on the other. The same twenty amino acids are used to make proteins in virtually all life forms.

Angiosperm: a flowering plant in the superclass Angiospermae that includes the order Legumiales of which the soybean is part.

Anther: the male generative organ of a plant.

Antigen: a foreign substance capable of eliciting an immunological response in a vertebrate animal, usually of a humoral variety and including production of an antibody specific to the antigen's structural makeup.

Artificial selection: the choosing by humans of a genotype that then contributes to the genetic types that make up the succeeding generations of a given organism or plant.

B

Bacillus: a genus of rod-shaped bacteria. *Bacillus thuriengiensis* is a spore forming soil bacillus that grows in the soils of many regions and is the source of the toxoid used in genetic engineering (see *Bt* toxoid).

Bacterium: a single celled, microscopic organism in the Prokaryote kingdom.

Base(s): the four structural subunits of DNA or RNA comprised of guanine, cytosine, adenine and thymine, or in the case of RNA, uracil.

Beta vulgaris: beets.

Bioaccumulate: the tendency for a substance to increase in concentration as it moves upwards in the food chain; a bioaccumulating chemical is usually soluble in fat.

Biogeographic realms: the divisions of the world's land masses according to their distinguishing animals or plants.

Bioengineering: the construction of a genetically controlled plant or animal by transferring genes from an otherwise genetically incompatible organism to create a novel function or product.

Biological species: groups of individuals which freely share a common set of genes and are reproductively isolated from each other so that interbreeding usually cannot occur.

Biomass: the effective weight of all the living materials in a given environment or ecosystem.

Blight: a fungal disease of plants often associated with wet conditions.

Bottleneck effect: a fluctuation in gene frequencies brought about by the abrupt contraction of a large population into a smaller one which then expands again with an altered gene pool.

Brassica: a genus of plants that includes broccoli and cabbage.

Breed: an artificial mating group derived from a common ancestor, often used for domestication.

Breeding: the controlled propagation of plants or animals.

Bromoxynil: a bromine-containing herbicide produced by Rhône-Poulenc Company under the brand name Buctril®.

Bt toxoid: the crystalline proteins derived from some strains of *B. thuringiensis* that is activated to become poisonous in the alkaline environment of an insect larvae's intestinal tract.

Bud: an underdeveloped plant shoot, consisting of a short stem bearing, overlapping immature leaves.

C

Cell: the smallest unit of all living things, capable of self-replication.

Chlorophyll: the green photosynthetic pigment carried inside plant cells in the chloroplast which permits the conversion of sunlight into chemical energy.

Chloroplast: the chlorophyll-containing photosynthesizing plant organelle thought to be descended from a bacterium known as cyanobacter which once colonized primitive plant cells. Chloroplasts contain their own DNA and can replicate independently of the cells they are in.

Chromosome: a threadlike structure made up of DNA strands and proteins (histones and non-histones) that carries genetic information in a linear sequence.

Coleoptera: the genus that contains the beetles.

Congener: in chemistry, one of a pair of structurally similar molecules or in plants, a member of the same genus or group.

Cotyledon: the leaf-forming part of the plant embryo in a seed. Once exposed to light, the cotyledon typically develops chlorophyll and

functions photosynthetically as the first leaf.
Crop lineage: the descendants of a single progenitor of a given food crop.
Cross: a mating between genetically different individuals of opposite sex or genotype.
Crossbreeding: see outbreeding.
Crystalloid: a form of protein which folds and stacks to form repetitive units, often with surface features of classic crystals, e.g., rhomboid, hexagonal, etc.
Cultivar: a variety of plant produced through selective breeding by humans and maintained by cultivation.

D

Darwinian evolution: the preferential reproduction of genetically varied organisms with specific adaptations that permit their differential survival.
DBHA: 3,5 dibromo-4-hydroxybenzoic acid; the breakdown product of bromoxynil.
Detritus: the disintegrated matter from inorganic weathering or organic decay.
Diploid: having a double genetic complement; the genetic material contributed from two haploid gametes.
DNA: the genetic material of cells comprised of bases, arrayed in an ascending/descending double helix.
DNA sequence: the linear array of bases (ATCG) which spells out the genetic code.
Deoxyribonucleic acid (see DNA): the molecular basis for heredity.
Dominant gene: a gene whose products are expressed when only one form of the gene is present as a single allele (See recessive gene).
Dose-response curve: the curve showing the relation between some biological response and the amount of a given toxin or unit of radiation received by the host.

E

Ecosystem: the composite of all the organisms and geographic features of a given place that makes up its defining characteristics.

Endosymbiotic cyanobacteria: an ancient bacterium thought to be the source of chloroplasts in plant cells.
Enzyme: a chemical which catalyzes biological reactions.
Epidemiology: the study of diseases among populations.
EPSPS: abbreviation for 5-enolpyruvylshikimate-3-phoshate synthase; a critical enzyme use by plants to make aromatic amino acids; the enzyme inhibited by glyphosate.
Essential amino acid: one of eight amino acids not synthesized in the human body, including phenylalanine; methionine; lysine; tryptophan; valine; leucine; isoleucine and threonine.
Estrogenic: having the properties of an estrogen, e.g., in stimulating cell growth or proliferation in specific sexual target tissues.
Eukaryotes: the superkingdom made up of organisms whose cells contain a true membrane-lined nucleus.

F

Flavonoids: molecules found in some plants that may have unpredicted biological properties, often antioxidant or hormonal in nature.
Flora: a flowering plant.

G

Gametes: the ova or sperm of animals; the pollen or ovules of plants.
Gene: the unit of heredity consisting of a sequence of DNA bases with "start" and "stop" information along with the base sequences for a specified protein.
Gene splicing: the creation of new genetic combinations by intentionally interspersing a novel gene sequence into an existing genome, usually in bacteria.
Genetic engineering: those experimental or industrial technologies used to alter the genome of a living cell so that it can produce more or different molecules than it is already programmed to make; also, the manipulation of genes to bypass normal or asexual reproduction.
Genetic information: the data contained in a sequence of bases in a molecule of DNA.
Genome: all of the genes carried by a given organism.

Germ cell: one of the sex cells which give rise to the gametes.
Germ plasm: the cellular lineage that gives rise to sperm or eggs.
Glycine max: the common soybean cultivated in the U.S.
GMO: abbrev. for Genetically Modified Organism; a plant or animal containing permanently altered genetic material.
Gossypium sp: cotton varieties.

H

Haploid: containing only half the normal complement of chromosomes; the genetic complement of the gametes.
Hepatocellular carcinoma: a malignant tumor of the liver.
Herbicide: a pesticide which usually affects only plants; a chemical with killing or growth inhibiting effects on plants.
Homogeneity: having the same form or content.
Homozygosity: having the same allele on both parental chromosomal strands; the state of being homozygous.
Hybrid: an organism derived from two distinct, and usually homozygous, parental lines.

I

Introgression: the introduction of genes from one member of the species into another where the donor is often geographically or morphologically distant from the recipient (see Introgressive hybridization).
Introgressive hybridization: the incorporation of genes of one species into the gene pool of another usually resulting in a population of individuals which continue to represent the more common parental line but which also possess some of the characteristics of the donor parent lineage.
Isogenic: having the same genetic makeup.

L

Lecithin: an emulsification product from soybeans used in chocolate and other foods.
Lepidoptera: a genus that includes the moths and butterflies; larvae of the same.

Locus: the site on a chromosome where a particular gene resides.

M

Mastitis: an inflammation of the udder commonly found in lactating dairy cows treated with rBGH.

Meiosis: the process by which a sexually reproducing organism reduces its chromosomes by half to provide gametes with a haploid genetic complement.

Metabolism: the process by which an organism processes food or other energy sources.

Metabolite: a breakdown product or sequential molecule derived from a particular molecule usually involved in one of the major chemical pathways in metabolism.

Microecosystems: the microbial colonies that comprise a given soil habitat or region (see microhabitats).

Microhabitats: the microscopically sized environments in which microorganisms live.

Monoculture: a growth or colony containing a single, pure genetic line of organisms; a genetically uniform line of plants or organisms derived from tissue culture.

Microflora: microorganisms that inhabit a given environment, including the human colon, or ecosystem; or soil microflora—the microorganisms found in soil.

Mitochondria: the energy-generating organelles contained in all higher eukaryote cells.

Molecule: an assembly of atoms into a structure maintained by inter-atomic bonds (e.g, hydrogen or carbon-carbon bonds).

Monocot: a plant whose seed has a single cotyledon; a grass of the monocotyledon lineage.

Monocotyledenous: having the properties of a monocot.

Monoxygenases: enzymes synthesized in the liver for detoxifying molecules in the body.

Monotonic: differing in a trait that changes smoothly over space or time.

N

Nitrile: a molecule containing a reduced nitrogen group.

Nitrogen fixation: the process by which atomic nitrogen is made accessible to plants by conversion to metabolizable chemicals like ammonia.

O

Opportunity principle: the concept in ethics that a large undertaking or commitment can limit the availability of resources for a like undertaking that relies on the same resources.

Organelle: a subcomponent of a cell, usually compartmentalized within its own membrane, e.g., a chloroplast.

Outbreeding: making crosses between distantly related members of the same species (Syn. outcrossing).

Ovule: the structure in seed plants which develops into a seed after the fertilization of an egg cell within it.

Oxydase: an enzyme which permits the addition of oxygen atoms to a given substrate.

P

Perennial: lasting for more than one year; plants that grow back from tubers or roots each season.

Phyla: the classification between Grade and Class; ex. the Arthropoda phylum (spiders, etc).

Pleiotropy: having many different effects from a single gene.

Pollination: the process by which the male sex cells of a plant's anther fertilize the stigma.

Propagation: the asexual reproduction and growth of plants from tissue culture, cuttings or scions from a parent plant.

Prokaryote: the superkingdom that contains forms of life lacking cell walls; the microorganisms lacking a membrane-bound nucleus containing chromosomes.

R

rBGH: abbreviation for recombinant bovine growth hormone, a genetically engineered stimulant of milk production.

Replicon: a self-replicating element which behaves autonomously during DNA replication; in bacteria, the chromosomal unit.

Rhizome: A rootlike, horizontal stem growing under or along the ground, and sending out roots from its lower surface, and leaves or shoots from its upper surface.

Roundup Ready™: the brand name for plants genetically engineered to be resistant to the herbicidal effects of glyphosate (Roundup®).

S

Seed bank: a collection of seed and germ plasm from a broad cross section of plants or food crops, in which seeds are stored under liquid nitrogen for protracted periods.

Shikamate synthase: the pathway inhibited by glyphosate.

Species: a freely interbreeding group of organisms that is genetically isolated from closely related stocks from which it might otherwise share genes; in classification, the individuals within an order that freely interbreed; (cf. genus and species: ex. *Glycine max*).

Stigma: the female portion of a plant, usually found at the tip of the style, into which the pollen tube grows.

T

Teratogenic: capable of producing birth defects or other reproductive harms manifesting in a visible disorder in form or size.

Transgene: a gene that has been moved across species lines into the germ line of a host.

Transgenesis: the science of interspecific movement of individual genes.

Transgenic: an adjective describing an organism that contains genes not native to its genetic makeup.

U

Utilitarianism: the ethical theory which uses cost/benefit analysis to maximize good over bad outcomes; "the greatest good for the greatest number" (cf. negative utilitarianism in which harms are minimized).

V

Vavilov, Nikolai: Russian geneticist and agronomist b. 1887, d. 1943; founder of the Vavilov Institute in St. Petersberg, which organized the first seed banking program.

Variety: a morphologically distinct subtype of a given species and genus; e.g., a novel variety of corn.

Z

Zea mays: corn.

Zygote: a fertilized egg of any diploid species.

Index